Praise for Ken Keffer's *Earth Almanac*

"Lively and lovely, Ken Keffer's *Earth Almanac* is a fun, enlightening guide to the wonders of nature throughout the seasons."

—*Foreword Reviews* (starred review)

"*Earth Almanac* makes a delightful daily read-aloud with family. Keffer's generalist approach offers encouragement to budding naturalists, inviting us to action as field data collectors and advocates for the earth."

—*BookPage* (starred review)

". . . a profusely illustrated and extraordinary compendium of facts and stories about the biodiversity and the natural world of North America."

—*Midwest Book Review*

"Keffer brings the relevance of the season to his descriptions with accuracy and charm. The illustrations, both full-color and line drawings, are wonderful and add a great deal to the enjoyment of reading *Earth Almanac*."

—*Booklist*

"In pointing out the curiosities and wonders of the natural world, Keffer invites you to suss out the relationships and connections between life forms and the environment they're situated in, both close to home and further afield."

—The Bookmonger

"Keffer has fun with his topics, and shares the bizarre and the everyday surprises all around us. . . . perfect for by the fire or on the back porch."

—Cool Green Science (The Nature Conservancy)

KNOWING THE TREES

Discover the Forest from Seed to Snag

Ken Keffer

Illustrations by Emily Walker

**MOUNTAINEERS
BOOKS**

MOUNTAINEERS BOOKS is dedicated to the exploration, preservation, and enjoyment of outdoor and wilderness areas.

1001 SW Klickitat Way, Suite 201, Seattle, WA 98134
800-553-4453, www.mountaineersbooks.org

Copyright © 2023 by Ken Keffer

Printed in China
Distributed in the United Kingdom by Cordee, www.cordee.co.uk

26 25 24 23   1 2 3 4 5

Copyeditor: Lori Hobkirk
Design and layout: Jen Grable
Cover illustrations and interior illustrations: Emily Walker

Library of Congress Cataloging-in-Publication Data is available at https://lccn.loc.gov/2023001120. The ebook record is available at https://lccn.loc.gov/2023001121.

Mountaineers Books titles may be purchased for corporate, educational, or other promotional sales, and our authors are available for a wide range of events. For information on special discounts or booking an author, contact our customer service at 800-553-4453 or mbooks@mountaineersbooks.org.

Printed on FSC®-certified materials

ISBN (hardcover): 978-1-68051-552-7
ISBN (ebook): 978-1-68051-553-4

FSC
www.fsc.org
MIX
Paper | Supporting responsible forestry
FSC® C008047

An independent nonprofit publisher since 1960

To Heather Ray—
Congrats on your book deal. I'm your
biggest fan. Love you always.

"The wonder is that we can see these trees and not wonder more."

—Ralph Waldo Emerson

Contents

The Life Cycle of a Tree

In the preface to *The Sibley Guide to Trees*, David Allen Sibley writes, "Trees are some of the most common things in our everyday lives and at the same time some of the most superlative. The tallest, heaviest, and oldest living things on earth are trees." Unlike Sibley's guide, which focuses on tree identification, *Knowing the Trees* inspires readers by imparting a new level of awareness and understanding of the natural world. Presenting the biome holistically offers an in-depth look at the forest, allowing you to see the forest *and* the trees. Most entries focus on trees that are native to and grow free in North America, while a few entries cover cultivated trees. But some naturalized species add to the confusion because, many times, these escaped ornamental plantings have flourished and even established substantial populations.

Despite growing up on the prairie, I have always felt that trees are synonymous with nature. Many other people feel the same way. Grasslands are wonderful biomes, and clearly trees alone do not represent all of nature. The connections between the two illustrate how drawn we are to the woods. Connections to nature are especially relevant and appealing in a world dominated by screentime and technology.

Our physical and emotional health is linked to the health of the earth. Forest bathing, for instance—a Japanese tradition that fuses mindfulness practices with nature—is gaining traction in the United States. As people turn to the forests for recreation and health, and for mindfulness, *Knowing the Trees* tells the stories of these woods.

According to the most recent Outdoor Foundation's *Outdoor Recreation Participation Topline Report*, "In 2018, 98 million Americans were considered moderate participants and spent at least ten days out of the year participating in outdoor recreation." The report claimed 47.9 million Americans took part in hiking, while nearly 42 million camped for at least one night. Adults who engage in these nature experiences usually do so two or three times per month. Despite the perception that people are too connected to electronic devices these days, many people clearly want to cultivate a connection to the outdoors.

This book celebrates one of North America's greatest biomes: the forests. From the coastal temperate rainforest of the Pacific Northwest to the iconic fall foliage of New England, *Knowing the Trees* highlights natural history, while weaving in the human experience with nature. Reading these pages, you will see forests with fresh eyes and get to know trees from seed to snag. Some of these amazing beings can live nearly five thousand years and reach staggering heights taller than the Statue of Liberty.

The life cycle of a tree serves as the framework for this book. Showcasing facts on the unique reproductive strategies of trees, the first section highlights the wonders of seed dispersal, shines a spotlight on some trees' remarkable ability to send up runners, and touches on grafting branches to create new trees. Seeds, which are important to trees for passing on their genetic materials, are also a bountiful resource for humans and wildlife.

The next section, "Roots, Buttresses & Knees," focuses on the foundations of trees. Roots offer stability while extracting essential nutrients. This vast underground network links individual trees and creates united forests, mysteries that this book unlocks. This section also examines buttresses and knees, the part of some trees' root systems that grow aboveground.

Tree trunks and the inner workings of the plants are the topics of the third section, the book's heart— or heartwood, as the case may be. Many forest products, from lumber and toilet paper to maple syrup, originate in the trunks. Functionally, tree bark protects the parts of the plant it surrounds. Nutrients and water flow through systems called the phloem and xylem, while hollow cavities shelter birds and mammals. It's fitting that this section is robust, as trees' woody stems separate them from other plants.

The treetop canopies are in an arms race for resources in the forest. Branches and crowns thrust leaves and needles outward to

gather light, and the shadowed understory is a by-product of this evolution. All forests are unique. Each one varies from the next in a multitude of ways. The species composition shifts as a forest matures. Even within relatively stable, late-successional climax forests, gaps in the canopy allow light to penetrate, creating a mosaic on the landscape. Section four shines a light on the photosynthesis happening in the treetops, or canopy, as well as on the contrast between deciduous leaf-bearing trees and needled conifers.

A tree's life cycle doesn't end with the death of an individual. Standing snags and coarse woody debris become essential habitats for forest fauna. Nutrients that have accumulated in trees over their lifetime are released into the environment at different rates— sometimes quickly because of a fire, and other times decaying slowly and feeding the decomposers. Section five highlights the afterlife of trees, including some of the most fascinating stories in the forest.

Because there is much more to a forest than the trees, the final section covers forest ecology. This broad topic considers the rest of the woods and the connections among its various species and highlights a diversity of forests, each with its own unique biotic and abiotic communities. It covers understory ferns in the northwest, along with Saguaro cactus and Joshua trees, none of which are technically trees but which function very much like forests. An entry on forest bathing reminds us that forest health is deeply related to human health. Throughout each life cycle

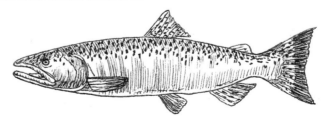

section, Seeds of Knowledge sidebars examine select topics and offer deeper insight into several matters unique to trees.

I hope that, armed with a more robust understanding of how forests function and how integral they are to human health, as well as to the health of our planet, your walk in the woods will be filled with a sense of wonder.

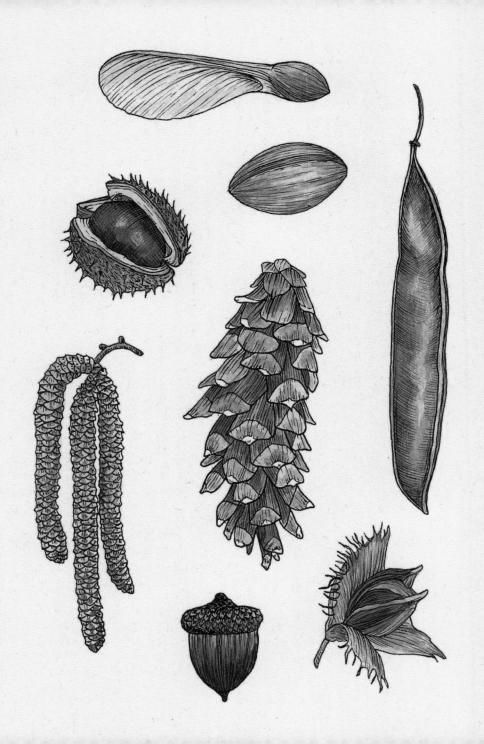

Cones
& Seeds

Knowing the Trees begins by broadly highlighting both gymnosperms and angiosperms. The gymnosperms (including conifers and ginkgoes) are the oldest lineages of trees, and these plants have exposed, naked seeds. By contrast, the angiosperms (deciduous trees such as maples and oaks) have seeds that develop within enclosed carpels, the female reproductive structures in flowers. Other differences between these two groups are more obvious to casual observers (needles versus leaves, for example), but this distinction between how seeds develop is at the core of evolutionary branching in plant life. It's a fine place to begin our journey through the life cycle of a tree.

Next, we'll discuss the birds and the bees of trees. It's easy enough to think of the seed as the beginning of a new tree, but where do these seeds come from? The sex lives of plants are especially complicated. Trees share genetic material in nearly as many ways as there are tree species. For some species of trees, males and females are separate individuals. Others can have both male and female flowers on the same plant, yet there are many mechanisms in place to prevent self-pollination.

Seeds are fundamental to the reproduction of trees, but they are also an essential source of food for many animals. Mast (or bumper) crops can sustain wildlife during peak years, but population crashes usually follow when crops don't produce as many seeds. A few tree species have outlived their seed dispersers, which you'll read about in Seeds of Knowledge: Evolutionary Anachronism, in this section.

This section closes with entries inspired by folk character and real-life arborist John Chapman. Nearly two hundred years after his death, the legends of John Chapman—aka Johnny Appleseed— live on. So do a few apple varieties he peddled throughout the eastern United States, although these days most spitter apples (small, usually tart varieties grown from seed) are about as fashionable as a tin hat.

Naked Seeds: Conifer Pollination

While people rave about the flowering bouquets that decorate the branches of many trees in spring, the stately conifers undergo much more subtle seasonal changes. One of the earliest lineages of trees, conifers don't depend on showy petals and sweet wafting aromas to entice pollinators. Instead, they rely on wind.

Most conifer species are monoecious, meaning individual trees grow both male and female parts. Although each tree has both staminate male and ovulate female cones, they also have adapted to keep from pollinating themselves. In many species, the timing is staggered, so that on any individual tree, the male and female cones mature at different times. Nearby trees will have slightly different rates of development, so neighboring trees can cross-pollinate each other.

Male cones are small and inconspicuous. In spring, these soft cones develop quickly and then scatter pollen spores with wild abandon before dropping to the forest floor by early summer. While it might seem like a nearly impossible feat for pollen spores to find receptive cones, if you've ever had the experience of yellow pine pollen blanketing every surface imaginable, you'll understand the vast quantities of pollen trees produce each spring. The male cones grow on lower branches, while the female cones are concentrated on higher branches. This separation of cone types minimizes self-pollination, as pollen spores aren't likely to land on branches directly above the cones that release them.

During this pollination stage, the female cones are also quite small. The female cone is fertilized when pollen lands on it. Seeds then begin to develop. In the gymnosperms (or "naked seeds"), the seeds grow at the base of the female cone scales. These cones take on the woody, familiar shape we often imagine when we think of conifer trees.

With Pacific yew—a dioecious conifer, one in which male and female flowers grow on separate trees—male and female trees can grow quite far apart from each other, posing a challenge for a pollination process that relies on the wind. Some isolated yews likely never pollinate another tree in their reproductive lifetimes. Once pollinated, female yew trees develop bright red arils, modified cone scales that resemble small, fleshy berries. Don't eat them—as with the rest of the yew plant, they are poisonous to people. Birds, however, can devour the arils. And even after passing through a bird's digestive tract, the seeds can still sprout.

Female ponderosa pine cone

Cherry blossoms

The Birds & the Bees of Flowering Trees

Perpetuating a species requires each generation to pass genetic information along to its offspring, and for trees, finding a suitable mate is a complicated task. Springtime, also known as allergy season, is when most tree species reproduce. In its basic form, pollination means transferring pollen from the anthers (male parts) to the stigma on the pistil (female parts). Pistils can be a single carpel or a collection of these structures. Angiosperms have seeds that develop in an enclosed carpel—which includes the stigma, style, and ovary. Evolutionarily, the angiosperms, including deciduous leafy trees, flowers, and grasses, are a much younger lineage than the gymnosperms, such as the conifers and the unique ginkgo.

Flowers can also be male (staminate) or female (pistillate or carpellate). Flowers that have both male and female parts are called perfect flowers. Adding to the confusion, both male and female flowers can appear on the same tree (monoecious, such as birches) or on separate trees (dioecious, such as holly, ginkgo, and willows).

Because they have both male and female flowers, self-pollination is an option for monoecious species, such as birch trees. But as with the conifers (see Naked Seeds: Conifer Pollination earlier), monoecious species rely on wind pollination to ensure fertilization takes place.

Showy, aromatic tree flowers are insect magnets. A proven, efficient way to maximize pollination of other relatively close-by neighbors, insect pollination is, however, a relatively late adaptation of the plant kingdom. The earliest evidence, preserved in a 99-million-year-old chunk of amber, has a tumbling, flowering beetle trapped inside.

Helicopter Seeds

Most plants are rooted in place—they can't simply get up and move when conditions change and make life difficult. There are no proverbial greener pastures in botany. Limited locomotion presents an added challenge when it comes time to disperse seeds. Suitable soil conditions, sunlight requirements, and water regimes are all barriers to the survival of seeds. Thriving adult trees may indicate growing conditions are great, but it's not ideal for seeds to drop directly below their branches. Mature trees can shade out the seedlings in the understory. That said, seeds rarely travel too far, and growing conditions may be substantially worse farther away.

Helicopter seeds in several species ensure that the next generation will maximize its potential for success. These plump seeds attach to paper-thin wings that help them land in just the right place—with or without the help of a child (or a youthful adult) tossing them around just to watch them fall.

A breeze can help propel the seeds away from the parent, but not so far away that the growing conditions will be drastically different. Botanically, the helicopter seeds found in conifers, such as pines and Douglas-fir, are different from those of deciduous species. In

Maple seed

the conifers, the wing is a pepper-like bract. Pines, firs, spruces, and hemlocks have seeds that flutter on single wings. The asymmetry of a lone propeller keeps seeds aloft nearly twice as long as double-edged bracts do. Having only a single wing means the seed takes longer to reach the ground.

The winged seeds of deciduous trees, called samaras, consist of a covered seed attached to the winged fruiting structure. Maples can have paired wings coming off each seed. The seeds of ash trees usually have just a single wing. Elm seeds are encased in wings that are nearly circular. On average, these symmetrical samaras don't flutter as far, but they still reseed the forests.

Synchronized Seeding

For some tree species, seed germination is either feast or famine. Drought conditions can have a negative influence on seed development and growth, while ample precipitation and sunlight can aid in the production of bumper seed crops. Early season cold snaps can damage buds and flowers, preventing trees from having seeds to drop in the fall. Well-timed rains can boost the growth of seeds, which is only the first step in tree development. Germination rates for seedling trees, like survival rates for most living things in nature, are astonishingly poor. Trees make hundreds or sometimes thousands of seeds at a time, but just-right-Goldilocks growing conditions are required for a tree to germinate.

Acorns from burr oaks

One approach is to release vast amounts of seeds but to do so at less frequent intervals. These mast (or bumper) crops of seeds occur every few years, although the mechanics of the process are still a bit of a mystery. Etymologically, "mast" originates from Middle English and Old High German and means *mete*, or food. This root word is also the source of the word "meat," which may help explain why the edible portion of nuts is sometimes called meat. It may seem like a stretch to think of deer and squirrels dining at an all-they-can-eat meat buffet, but that's basically what a mast crop is. Trees are likely expending stored-up nutrients during these peak mast years, and afterward, it takes them a few years to rebound. How an entire stand of trees synchronizes its seed releases, however, still stumps scientists. It could be as simple as nearby trees experiencing similar weather patterns. Maybe they are sharing resources via mycorrhizal fungi (see Seeds of Knowledge: Underground Web, in Roots, Buttresses & Knees). Or perhaps the trees are reacting to chemical signals released by adjacent trees.

Oaks are a classic example of mast crops, but plenty of other nut-producing trees, including hickories, walnuts, and beeches, fall under a hard mast umbrella. Over time, you may notice trends in the woods you frequent. Some years in fall, with so many nuts covering the forest floor, you may feel as if you are walking on marbles. Other years, you may be hard pressed to find a single seed. This phenomenon can happen with soft mast trees as well, including hawthorn, dogwood, and mulberry, but the peaks are less extreme.

In what is basically a predator-prey relationship, seed eaters will gorge themselves when seed resources are available. Many animals of all sizes from mice to bears feed on these nutrient-dense foods. Seed predators help distribute and disperse nuts and berries throughout the forest floor, and enough seeds will be spared to ensure that they germinate and the forests continue. As animal populations cycle through boom and bust years, mammal numbers can increase following solid mast years in forests.

Cache Stash Middens

While turkeys, deer, bears, mice, and many other species capitalize on a bumper, or mast, crop of nuts by gorging on them, squirrels, jays, crows, and others take matters into their own hands—er, paws and beaks—and hide morsels for later. Hoarding of food resources isn't unheard of in the animal kingdom, and it benefits plants too.

Tucking edible bits away for future snacking is a behavior known as scatter hoarding. Most squirrel species bury seeds and nuts far and wide. This isn't random. White oak acorns germinate in the fall, and squirrels consume these nuts more readily. Red oaks germinate in the spring, and so the rodents are more likely to scatter hoard them. Similarly, jays, nutcrackers, woodpeckers, and chickadees distribute seeds widely as they stock up nature's pantry for the upcoming winter season. The animals won't recover and consume all these seeds, and some of the ones left over will germinate and grow where they were inadvertently planted.

Red squirrels have a different approach to hoarding. Instead of scatter hoarding, these cone specialists, sometimes called pine squirrels or chickarees, cache a collection of food in a single place. This midden of cone scales accumulates as the squirrels extract and eat the conifer seeds. Midden piles can stretch for 15 feet and often surround the base of a preferred eating stump. If you come anywhere near these piles, chattering chickarees may scold you for intruding on their stash.

Acorn woodpeckers store up to fifty thousand seeds in a single granary pole. Family units of these woodpeckers tend to a territory

Acorn woodpecker on a granary tree

cooperatively, using these trees for many generations. These birds tap acorns and other nuts into crevices and holes. As the seeds dry, they shrink, and the woodpeckers move them into smaller holes.

Edible Seeds

Tree seeds aren't food for just woodland critters. People also eat a variety of tree parts. Sometimes we refer to them as seeds, while at other times we call them nuts or fruit. With fruit, humans prefer the fleshy bits that surround the seed and often discard the seed proper; however, you cannot always assume that seeds or nuts are all "fruit." Botanically, these fun feasts are from a variety of plant structures.

To a botanist, a fruit is the structure that develops around a seed, and it derives from the ovary of the flower. Yet not all fleshy orbs are created equal. Apples are a fine example of a true fruit: they have a bunch of seeds surrounded by soft, tasty bites. The pawpaw flesh also surrounds several seeds, but unlike apples, the pawpaw is a prized wild fruit that doesn't flourish when cultivated. The soft pulp of the pawpaw, also called the *banana of the Midwest*, has a sweet mango-like flavor.

One instance when fruit is not a fruit is when it's a drupe. While fruits develop around multiple seeds, drupes contain single seeds protected by hard casings. In the kitchen, we often call this hard casing the pit. Drupes are sometimes referred to as stone fruits, such as plums, peaches, and cherries. Almonds, cashews, and pistachios are

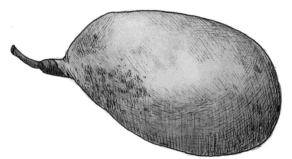

also drupes, although in these instances, the prized bite is the seed, not the flesh.

Black walnuts are drupes too. These racquetball-sized, but tennis-ball-colored drupes are notoriously difficult to crack open. Wildlife eat these seeds. Although some people forage for black walnuts to eat them, they are more frequently used as a natural dye. Artists have used ink created from black walnuts for centuries, and walnut dye turns fabric a rich brown.

Pawpaw fruit

You may ask: What are nuts, then? Technically, most tree nuts are in fact a type of fruit. Nuts are the seeds of trees protected within dry shells. These shells come from the ovary of the plant and are, therefore, functionally fruits. Each nut contains a single seed. For example, acorns are the seeds, or fruits, of oak trees. Other examples include chestnuts, pecans, and hazelnuts.

Pesto recipes usually call for pine nuts, the delicious seeds of pine trees. Pine nuts, however, lack the hard outer shell of true botanical nuts. While there are dozens of potential sources of pine nuts, most sold commercially are harvested from Korean pines grown in Russia. The piñon pine that grows in the American Southwest contributes roughly 20 percent of the piñon pine nuts on the market.

Crossbilled Cone Robbers

Producing enough seeds to survive and regenerate the forest requires trees to perform a balancing act. For every tree seedling that survives, many more are sacrificed. Conifers, such as pines, spruces, hemlocks, and firs, bear cones. Although cones protect conifer seeds while they develop, as the seeds mature, it is beneficial for the plant to distribute them far and wide. Crossbills, including red- and white-winged, are especially well suited for exploiting this bounty.

Few species of birds have a stronger connection to conifer trees than the crossbills. The mandibles of their beaks overlap along the thin tips. These offset beaks allow the birds to pry into cones and extract the seed nuggets. Crossbills are also one of the only birds that

Red crossbill

can stick out their tongue. The birds use their crossed bills to nibble apart the cone scales, followed by a flick of their tongue to extract the seed snacks.

Roughly 75 percent of the white-winged crossbill species have lower mandibles that twist to the right, providing them with the best angle for foraging. These crossbills predominately feed on spruce and tamarack cones. The birds snip the cones from the branches before dining. They use their left foot to clamp down on the cone while nibbling on the seeds to the right.

For red crossbills, the ratio of left- and right-crossed bills is closer to 50/50. Red crossbills rarely grasp the cones when feeding. Instead, this species usually eats from cones that remain attached to tree limbs. The birds can't always reach what they are trying to get at when foraging. The direction of their crossed beak limits access, so they end up abandoning several seeds in the cones. Right-crossed crossbills have access to some seeds, while left-crossed ones have access to other seeds. Despite coming at the cones from both directions, crossbills still leave many seeds behind, which adds to the pool of seeds that may germinate someday and become part of the forest.

Ornithologists have identified many different red crossbills, based on bill sizes and vocalizations. In most instances, these distinctions represent different types, not unique species. For example, Type 1 refers to the Appalachian crossbill. This medium-billed bird eats from a variety of cones and is found in the eastern part of the United States. A smaller-billed Type 3 red crossbill is primarily a hemlock-feeding specialist in the northwest. Formerly known as the South Hills crossbill or Type 9, the large-billed Cassia crossbill became recognized as a separate species in 2017. Although sometimes the various types can breed, other red crossbill types might be recognized as individual species.

The Cassia crossbill of southern Idaho evolved in lodgepole pine forests without competition from red squirrels. These less nomadic crossbills appear not to breed with other types of crossbills, even when their ranges overlap. Isolation makes the Cassia crossbill susceptible to threats and extinction. A 2020 fire burned through much of the Cassia crossbill range, and some researchers estimate 40 percent of the entire population was affected. A warming climate and regional wildfires may wipe out the species entirely someday.

Most crossbills are noted for epic irruptive, or migratory, movements. For many crossbills, migration isn't simply a matter of flying south for the winter. Instead, these birds are highly nomadic. Some years, crossbills hardly move around on the landscape, but at other times, the birds roam across the continent in search of food. Cassia crossbills are the exception as they hardly roam at all. What is especially remarkable is that crossbill nesting is also tied to cone crops. Crossbills can lay eggs and raise their young even in the dead of winter if they have access to enough seeds.

Although there may be years between sightings in a location, when they do visit an area, crossbills make themselves known. Large flocks can descend on stands of conifers. With some of the oldest trees in a community, cemeteries can make for solid crossbill watching. Many types of crossbills will also eat from tray feeders filled with sunflower seeds, so as you wander through neighborhoods, keep your eyes peeled for these unique seed eaters.

Serotinous Cones

After a fire, a landscape can look bleak. An unfathomable amount of forest biomass has gone up in smoke, with only a few charred timbers standing tall, like gnarled, blackened sentinels looking over the disturbance. Yet, for many trees, this natural process is key to releasing the seeds that will become the next generation. Species such as lodgepole pine, Jack pine, and giant sequoia have specialized serotinous cones that require the heat of a fire to release their seeds.

Conifers (pines, spruces, and hemlocks) have both male and female cones. The male versions are where pollen originates. These small, soft parts are easy to overlook on the tree branches, unless they are releasing allergy-inducing powder into the air. When you think of a typical cone, you're likely imagining female cones, woodlike structures composed of several scales that form a familiar bundle. These female cones protect the seeds that develop at the base of its scales from adverse weather. Scales often have sharp tips for added protection from seed eaters such as birds and rodents. Seeds in many cones are typically released each autumn. However, seeds remain in fire-adapted serotinous cones for years at a time.

A layer of resin acts like a sealant that keeps serotinous cones locked up—a natural storage unit for the developed seeds—until the heat from a fire unlocks the cone. The seeds often remain protected, even as the serotinous cones open initially, but rainfall will eventually dislodge them and also provides the moisture required to spur new trees to start growing across a burned area where forests once stood.

Serotiny is most associated with fires and conifer cones, but the term also applies to various triggering mechanisms that stimulate seeds to release, not simply at seed maturity, such as the death of the parental tree, moisture, or dry conditions. It may be hard to imagine trees as nurturing beings, but by releasing massive amounts of seeds during ideal conditions, a tree is improving the odds that some of its seeds will survive and ensuring its genetics are passed on to another generation.

The effects of the 1988 fires that burned Yellowstone National Park are still very evident on the landscape. Decades after those burns, the area is slowly returning to woods. The fires merely reset the successional process. Following the fires, the park's growing conditions became ecologically suitable for seed germination: nutrients were released in the soils and competition decreased. Fireweed and other plants established themselves in the short term, as is true in all forests following a fire. These early successional plants stabilize soils so other seedlings and young trees can reclaim the land.

Lodgepole pine cone

Research conducted in Yellowstone shows that lodgepole pines have two cone adaptation strategies. Before the extensive fires of 1988, serotinous cones were more common in lower-elevation lodgepole forests. The fires released this cache of seeds, and these lower-elevation lodgepole forests have grown back in thicker doghair, or young, stands following the burns. Historically, fire frequency in these zones ranged from 135 to 185 years. Higher-elevation lodgepole forests, however, had a longer duration between fires—280 to 310 years on average—and these forests regenerated with sparser stands of trees. It turns out that these higher stands of trees relied more heavily on traditional cones releasing seeds annually.

Lodgepole pine forests outside Yellowstone that also experience fire less frequently exhibit a lower prevalence of serotinous cones. When a stand of trees in Montana was affected by fire, more than 75 percent of the cones in the regenerating forest were serotinous. In nearby lodgepole forests disturbed by something other than fire, the serotinous rate was closer to 25 percent of cones in new trees. Serotiny also seems to increase as lodgepole pines age. Human efforts to suppress fire have shifted the effects, altering ecosystems, but on an evolutionary scale, historical fire patterns help explain the prevalence of serotinous cones in a forest.

One-of-a-Kind Ginkgo

When it comes to trees, none is as unique as the ginkgo. This plant's evolutionary fossil history stretches back in time roughly 270 million years and represents one of the few broadleaf deciduous gymnosperm trees. Others in the earliest tree lineages are the needle-bearing conifers. Only a few types of ginkgoes appear in the fossil record, and the ginkgo that exists today is the only species categorized in an entire division of the Plantae kingdom, meaning it has no close, or even distant, similar relatives.

Likely now extinct in the wild, the ginkgo was probably cultivated in China one thousand years ago. In the late 1600s, Engelbert Kaempfer of the Dutch East India Company mentioned and illustrated ginkgo from southern Japan. Less than a century later, the species was being cultivated in parts of Europe and North America. It is often touted as an ingredient in medicinal treatments, with purported benefits ranging from boosting brain function and memory to improving circulation and heart health, although its effectiveness is debatable in most instances.

These days, ginkgoes are ornamental and distributed in pockets around the world. It has a moderate tolerance for many growing conditions but doesn't do well in especially hot, dry environments. The species has both male and female trees. One unique aspect of pollination is that ginkgo trees produce motile sperm. In the spring, male trees release pollen grains containing the sperm. When these particles land on female ginkgo trees, pollen tubes develop, and roughly four months after the pollen is initially dispersed, sperm, aided by thousands of flagella tails, swim to an egg to try to fertilize it. This active sperm is a relic trait that stems from when aquatic vegetation colonized land 475 million years ago. In addition to ginkgo trees, motile sperm is also found in cycads, ferns, horsetails, and mosses.

Ginkgo leaves

A fleshy yellow coating surrounds the ginkgo seed, and as the fruit ripens, it's hard not to notice the powerful stench, which earns the tree its well-deserved reputation of being stinky. The aroma likely attracted dinosaurs, which helped disperse the seeds. The stench doesn't keep squirrels away, however, and these days small rodents play a key role in helping ginkgoes disperse.

Ginkgo leaves grow in a distinctive fan shape. The venation, or arrangement of veins, is closer to that of ferns than to other kinds of trees. Its canopy turns bright yellow in fall, but this brilliance is often short-lived because the leaves fall nearly in unison. Ginkgoes can reach heights of 50 feet, growing quicker in full-sunlight conditions. Its hardy nature allows it to grow easily, live a long time, and resist pollution, insect damage, disease, and fungal infection. Four individual ginkgoes growing one mile from Hiroshima, Japan, even survived the atomic blast in August 1945.

The Gastronomical Piñon-Juniper

The piñon-juniper habitat of the American Southwest covers roughly 15 percent of the Four Corners states—Utah, Colorado, Arizona, and New Mexico—plus adjacent Nevada. Species composition varies in this forest, which extends from lower-elevation grasslands and shrublands to higher-elevation ponderosa or montane forest types,

Gnarled old juniper

but it is dominated by piñon pines and junipers. Prime conditions are between 4,500 and 7,500 feet in elevation.

By excavating bushy-tailed woodrat (packrat) middens and examining pollen cores, researchers have determined the piñon-juniper ecosystem dates back to the Wisconsin glaciation 10,000 to 15,000 years ago. As climate conditions change, the entire landscape continues to shift. Warmer, drier conditions force piñon-juniper woods farther north and to higher elevations.

If you give a juniper berry a strong pinch—really crunching it between your fingers—you may recognize the smell. These berries are a base for gin. In *The Tree Forager*, Adele Nozedar observes, "And gin is still incredibly popular these days—in far more salubrious watering holes than in centuries past you can buy just about any flavour or colour that you fancy. It's great fun." Harvested mostly in Eastern Europe, juniper berries aren't fruits but modified seed cones. As a component of cocktails, juniper adds a flavor that hints at being slightly medicinal. The berries are also used in the culinary realm, where they are often paired with game meats, including duck.

Many pine nuts are harvested from the state tree of New Mexico, the two-needled piñon. These pine nuts are often used in pesto recipes. In fact, the American Southwest produces 20 percent of the world's supply of pine nuts besides those that come from Europe and Asia. In the Four Corners region, people reluctantly share their harvests with Steller's jays and Clark's nutcrackers.

Paleoclimatology

Seasonal allergies are a reminder that the smallest things can affect us in significant ways. Collectively, microscopic spores from plants, shrubs, and trees can blanket a landscape, but individually, they reveal events that occurred in ancient history. The largest grains are less than half a millimeter across, and hundreds of the smaller ones would fit on a pinhead. The sporopollenin that makes up the cell walls of pollen grains is chemically stable. Each species can be identified by the unique shape of its pollen, and because the grains don't break down easily, researchers can map out the flora of the past by mining for spores.

Most pollen samples are collected from cores drilled into sediments at the bottom of ponds, lakes, or even the ocean floor. Pollen settles into the muck, and year after year, this pollen-infused mixture accumulates like layers on a cake. When cross-referenced with other data, such as radiocarbon dating, paleoclimatology scientists can document past climate trends. The pollen record also reflects how the surrounding land has changed. As the eastern forests of North America were cleared, ragweed pollen ticked upward while spores from woody species declined. Pollen dating is a great tool for reconstructing the past, but it has limitations in arid climates.

Another technique for examining the vegetation of the past is to unearth middens made by bushy-tailed woodrats, or packrats. These piles of debris include vegetation, packrat waste matter, and anything else nature's hoarders collect. Researchers at the University of Arizona's Desert Laboratory pioneered this method and have been using it for decades. So far, they have analyzed more than two thousand fossilized middens from western North America. Parts of plants in these middens were pickled in packrat urine. These crystallized masses, cleverly termed amberat, preserve organic remains of plants and animals, allowing researchers to reconstruct the past. Packrats have relatively small ranges, rarely roaming more than 100 meters, so each sample offers a snapshot of local conditions and trends. For example, some amberat contains plant debris suggesting a shift from wetter environments to more drought-tolerant plants, which is evidence over time that the local climate is changing.

Extensive woodlands characterized most Pleistocene vegetation in what is now the American southwest, including spruce-fir, mixed conifer, and subalpine forests. In the last ten thousand years, these forests have receded northward to the highest of elevations and have been replaced by stands of piñon junipers. Scrublands characterized by creosote have expanded north from Mexico into what we now recognize as the Chihuahuan, Sonoran, and Mojave deserts.

Bushy-tailed woodrat

Evolutionary Anachronism

Animals distribute plenty of plant seeds. Trees with calorie-rich, tasty, ready-to-be-consumed fruits and nuts entice many birds and mammals. The term "endozo-ochory" describes the process of seeds passing through animal digestive tracts unharmed and then being dispersed on the forest floor. Sometimes, this scarification process weakens the protective layer on the seeds, which helps the passed seeds germinate more easily. Bears, for example, play a role in the germination of many fruit-bearing plants. Natural resource managers have planted bear scats, and these "poop plants" have grown at better rates than seeds that haven't taken a trip through ursid intestines.

Curiously, researchers have discovered that a few tree species seem to have lost their primary seed partners. Connie Barlow highlights much of this evolutionary anachronism work in *The Ghosts of Evolution*. Dan Janzen and Paul Martin first pondered the puzzle of germination with accumulated piles of fruit in the tropics, but there are examples from North America too.

Osage-orange is a classic species without an effective modern wild seed eater. This relative of breadfruit and jackfruit produces softball-sized fruits, sometimes called hedge apples, fleshy yellow spheres that look like small heads of chartreuse cauliflower. For a plant that devotes so many resources to creating fruit, there has got to be an evolutionary payoff in the form of seed dispersal. The pulp was likely a favorite of mastodons, an extinct member of the elephant family. Mastodons' spiked teeth were perfect for noshing on Osage-orange fruits, and like other mammalian seed dispersers, they would travel far from a parental plant before spreading the seed in a pile of mast-o-dung. According to the pollen record, the distribution of Osage-orange became very restricted after mastodons went extinct. The fruit on Osage-orange trees matures at roughly the same time, so these mature hedge apples all hit the ground within a few days. A few may roll down gentle slopes, but the overall range will not expand like it did when it was spread by mastodons. Nowadays, people and hungry horses have replaced mastodon dispersal of Osage-orange. The tree is cultivated as fencerows throughout much of the Midwest, giving it a spotty but widespread distribution. When pruned, it sends out dense suckers that create thickset thorny hedge lines, nature's barbed wire. People don't generally eat hedge apples, but some folks use them as a measure to control pests, similar to how mothballs are used—a practice whose effectiveness seems limited.

The honey locust is another evolutionary anachronism. Intimidating thorns on its trunk serve as a physical barrier that keeps plenty of potential browsers at bay. Its seeds are encased in an 8-inch leathery, but sweet-fleshed, pod. Both the seeds and pods have been found in preserved mastodon scats. Again, the mastodon would have been able to use its sharp molars to easily pierce and grind open the pods, aiding in the distribution of the seeds.

Fruit of an Osage-orange,
aka a hedge apple

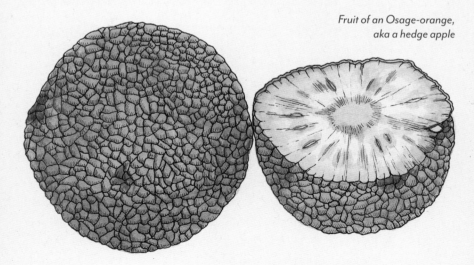

Thick pods similar to the honey locust's protect the seeds of the Kentucky coffeetree. Pawpaw and persimmon also have oversized fleshy fruits that would have benefited from Pleistocene plant eaters. The seeds of all these trees can still germinate without these Pleistocene herbivores, yet the loss of these animals has seemingly affected the rate and distance of seed distribution. Many modern animals eat the flesh, but none can replace these extinct seed distributors. These days, in the absence of endozoochory, the trees rely on vegetative regeneration as much as on seed germination. Prolific root suckers have likely sustained these plants in the thirteen thousand years since their megafauna foragers died out.

Johnny Appleseed

As with many folk heroes, what we know about Johnny Appleseed has become a combination of truth and legend. John Chapman was born September 26, 1774, in Leominster, Massachusetts, and died in Fort Wayne, Indiana, seventy-one years later. The details of his life, however, along with the exact date he died in 1845, remain a mystery.

Land records are one of the more tangible pieces of evidence to examine when piecing together the biography of Johnny Appleseed. You see, he didn't traipse around in the woods, sprinkling apple seeds, as most of us believe; instead, Mr. Chapman was a methodical businessman, accumulating wealth in the form of planted, and planned, orchards. In the United States in the late 1700s and early 1800s, people could claim property by clearing and planting the land. Johnny Appleseed developed thousands of acres of fruit orchards to claim that property as his own. While he sold off many of these orchards in his lifetime, at the time of his death, he had more than 1,200 acres to his name across Pennsylvania, Ohio, Indiana, and Illinois.

Red Delicious and Golden Delicious apple varieties are likely linked to Appleseed's plantings, but most crops harvested weren't intended to be snacks. Small, tart apple varieties, collectively referred to as *spitters*, were the norm in the 1800s, and they were turned into vinegar, cider, and applejack.

A Red Delicious apple

He is often depicted as a generous man who lived a life of self-imposed poverty as he wandered around barefoot in his clothes made of coffee sacks. Beyond spreading apple seeds, he also spread a religious message as a follower of the Swedenborgian Church. Whether he wore a tin pot on his head as a hat is debatable. A long-handled pot likely wasn't one of his fashion accessories, but perhaps he covered his noggin with a metal bowl. The Fort Wayne TinCaps minor league baseball team certainly supports the notion. Its logo is a caricature of an apple as a person's face, wearing a tin pot upside-down and backward, like someone would wear a baseball cap.

The only known surviving Johnny Appleseed tree grows in Nova, Ohio. Ironically, the plant survives thanks to grafting, a cultivation practice of splicing branches onto new trees, that John Chapman staunchly opposed because, according to his religious beliefs, the practice caused the trees to suffer. More than ten thousand genetic clones of this tree have been distributed far and wide from the original.

Grafting

You'd think chopping down a tree would be a fatal blow, but much like human skin can be transplanted via a medical procedure in humans, it is possible under certain conditions to splice trees and create a hardier species. Although quite rare, natural grafting has been documented in more than 150 species.

Grafted tree branches

People first harnessed the technique of grafting plants thousands of years ago. The "father of horticulture," Theophrastus, was a student of Aristotle's and wrote about grafting in 300 BC, though the technique was probably used long before that. Many species can grow from cuttings that resprout, but grafting takes this process of regeneration to the next level. Grafting essentially involves cutting a twig from one tree and planting it onto another tree. The key to success is ensuring that the vascular cambium layers beneath the bark connect and fuse. The phloem contact point can heal in as little as three days, allowing food to flow, while the xylem linkage that transports water may take a week or longer. The stem to be cultivated is termed a scion, and it can grow on the rootstock of a different species.

Grafting has many benefits in the horticulture realm. Apple trees grown from seeds—or spitters—are notoriously rogue, producing fruits that are often vastly different from the seeds they sprout from. Henry David Thoreau once noted that spitter apples can be "sour enough to set a squirrel's teeth on edge and make a jay scream."

Grafting can harness the genetic variability of apples. This crop consistency creates reliable products. Similarly, the hybridization of other fruit crops can be maintained with grafting.

For conservationists, one drawback to grafting can be a lack of genetic diversity as opposed to reproduction and regeneration from naturally dispersed seeds. On the other hand, to disperse isolated populations or combat hybridization or disease, grafting techniques are implemented to preserve genetics. For example, botanists are splicing American and Chinese chestnuts to balance genetic preservation with disease resistance. Individual butternut trees with pure genetic strains are selected for grafting to avoid losing the species to hybridization. Similarly, in China and Guam, grafting is used to maintain species with limited distributions and populations that are susceptible to being wiped out in a single natural disaster.

Orchard National Park

While there is no Orchard National Park in the United States, there very well could be. More than one hundred properties of the National Park Service system, including units as unique as national monuments and seashores and national historic sites and battlefields, have living fruit trees that are at least fifty years old. National lakeshores and national parks are also orchard hotspots. Always check with local authorities about regulations, but sometimes harvesting fruits is legal and even encouraged within a handful of NPS properties.

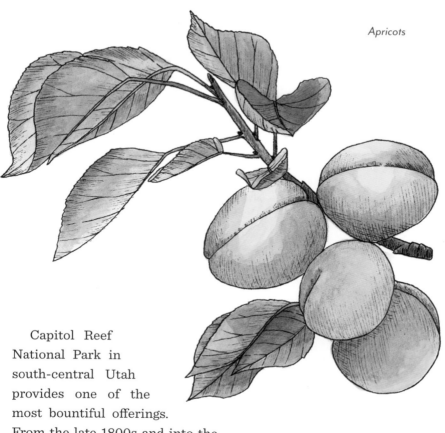

Apricots

Capitol Reef National Park in south-central Utah provides one of the most bountiful offerings. From the late 1800s and into the early twentieth century, acres of fruit trees were planted along the banks of the Fremont River and Sulphur Creek. When the community of Junction sprang up in a valley in western Colorado, so did thousands of apricot, cherry, plum, and apple trees. By 1902, the settlement had become known as Fruita. Gravity-fed flood irrigation ditches allowed crops to flourish in an otherwise dry landscape.

These days the National Park Service manages the orchards, which are listed on the National Register of Historic Places as the Fruita Rural Historic District. The heirloom orchard grounds are open to park visitors. The u-pick seasons range from approximately June through July for cherries and apricots, August and September for peaches and pears, to mid-October for apples. Capitol Reef National Park provides ladders and picking tools, and visitors pay a small fee at a weigh station for the fruit they harvest.

In addition to its orchards, the national park is also a destination for geology enthusiasts. The region, named for the landscape's resemblance to the domes of many capitol buildings, showcases the Waterpocket Fold. The rugged cliffs and domes of this monoclinal fold stretch for roughly one hundred miles.

Happy Holly-Days

Worldwide, there are more than five hundred species in the holly family. With this many types, you can imagine the range and diversity of niches the hollies fill. Some holly trees can reach a height of 80 feet and maintain thick, leathery green leaves throughout the year. Other hollies are vinelike. Many drop their leaves annually. Some species are monoecious, meaning that the same plant has both male and female parts. Others are diecious, where only female trees sprout the berrylike drupes. Although red is the color most strongly associated with holly berries, some species sport fruits closer to orange or yellow in tone.

There are fifteen different hollies in the United States. Besides the American holly, other related plants include winterberry, dahoon,

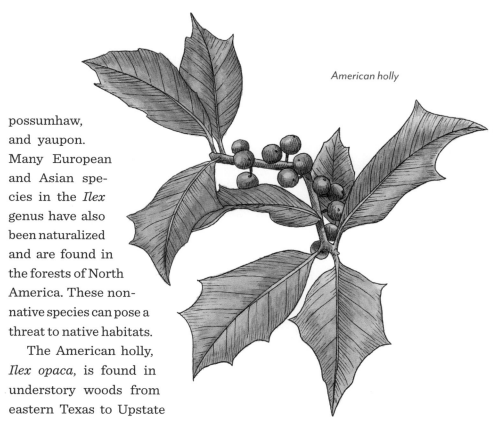

American holly

possumhaw, and yaupon. Many European and Asian species in the *Ilex* genus have also been naturalized and are found in the forests of North America. These non-native species can pose a threat to native habitats.

The American holly, *Ilex opaca*, is found in understory woods from eastern Texas to Upstate New York and coastal Massachusetts. It was adopted as the state tree of Delaware in 1939. It's a popular choice for winter and holiday decorating because of its hardy green foliage and vibrant late-season berries. Historically, Delaware was a major exporter of the plant, although artificial versions have diminished the demand for fresh boughs.

The fruits of holly are mildly toxic to humans and pets, but a few bird species can tolerate ingesting them, so holly can be an important winter food source for waxwings, mockingbirds, and thrushes, including robins and bluebirds. Herbivores also nibble occasionally on holly leaves, which triggers the tree to grow replacement leaves that are extra prickly.

Roots, Buttresses & Knees

Roots can be easy to overlook since, for the most part, they hide out underground. They function kind of like a skeletal system, as well as a version of a tree's circulatory and digestive systems. Being rooted in place can seem like a disadvantage for a living organism, yet a solid foundation is vital to keep a tree from toppling over.

Roots link a tree not just to the soil but to a vast network of mycorrhizal fungi. Research into the nutrient sharing and hoarding between roots and fungi is a fascinating topic that people are only beginning to understand. Individual trees appear to collaborate with each other. This support is often focused on genetically related individuals, but different species will occasionally join forces across this fungal association.

Sometimes, competition plays out on the ground level with allelopathic defense mechanisms in place. These chemical compounds secreted by plants can stunt the growth of, or even kill, nearby plants by essentially poisoning their root systems.

A few species, such as cypress and mangrove, are especially known for their roots. These unique beings get their own entries in this section, and you will gain an appreciation for their specialized, wet roots.

You will also discover the wonders of the banyan tree in Lahaina, Hawaii, and the extensive quaking aspen tree known as the Pando grove in Utah. The banyan—a single tree that takes up an entire city block—is a cultural icon. And the Pando grove is perhaps the largest living organism on the planet. The roots on the 100-plus acre tree are roughly eighty thousand years old, while the individual sprouted tree trunk stems are often younger than one hundred years old.

Roots provide resources to the rest of the tree, but they are also vulnerable to diseases, such as types of root rot. Many species emphasize root growth in the early years of plant development, an approach that longleaf pines take to an extreme. The plant resembles individual stems of grass sprouting until it has established a well-structured root system. Then after a couple hundred years, this tall timber grows to its full potential.

Busting Root Myths

It is bad enough that tree roots are often overlooked, but even worse than being forgotten, root systems are subject to several misconceptions and outright fallacies. One common myth is that the root system of a tree is basically an inverted mirror image of the canopy, leading to the false conclusion that the roots rarely extend beyond the drip line, or the width of the canopy. While the extensive

Sucker shoots sprouting from a tree stump

network made up of branches, limbs, and twigs is reminiscent of the underground lumber, they function in different ways. Leaves jockey for position to gather sunlight. Roots, on the other hand, radiate outward, expanding their reach as they mine the soil for water and nutrients.

Another misconception is that tree roots are responsible for underground damage to sewer lines and home foundations. There is no denying that roots can infiltrate these nooks and crannies. However, costly home repairs shouldn't be blamed on roots alone. It's much more likely that shifting soils caused the initial damage. Tree roots are hardly innocent bystanders, though, as the fine-tipped roots will sneak into the slightest of openings. It is important to plan for root growth when undertaking any new landscaping or home remodeling projects.

Another myth is that if you cut down a tree, the plant dies. This fact holds true for some species, but in many instances, the roots can resurrect the trunk. You've likely seen an old tree stump with many sucker shoots sprouting from what should logically be a lifeless chunk of wood. These sprouts are how the Pando grove of aspens in Utah (see Pando Grove, later in this section) has remained alive for so long.

This potential for regrowth is also the bane of people trying to eradicate invasive plants from the landscape. If a tree of heaven is cut down, for example, new shoots can spring up some 100 feet away. It is necessary to treat stumps with herbicide to kill off some of these persistent invaders.

Root Systems

Much like the tip of an iceberg, the part of the tree visible above-ground isn't the complete picture. What goes on underground is crucial. The root system of a plant serves multiple functions, including water and nutrient gathering, stabilization, and support. A rough estimate of how far roots spread can be obtained by measuring the tree's diameter at breast height (DBH) in inches and then converting to feet. For consistency, foresters measure DBH at 4.5 feet above the ground. The tree's diameter in inches at that spot gives you an idea of how many feet the roots will spread. For example, a 20-inch DBH means roots can extend 20 feet out from the tree trunk.

There are three main types of roots—lateral, tap, and oblique or heart—and many plants have a combination. Their growth is based on both genetics and soil conditions. Extending far from the trunk but growing at relatively shallow depths, lateral roots dominate roughly 80 percent of tree species. Deciduous trees such as maples, cottonwoods, and ashes are all examples of species with lateral roots. When exploring a forest, you may find toppled trees with exposed gnarls of lateral roots. Eroded riverside cutbanks provide another opportunity to glimpse the world of lateral roots that are normally underground.

For trees such as sweet gum, hickories, tupelo, pines, and many of the oaks, root systems are defined by a strong taproot that augers below the trunk. As trees age, the taproot anchor becomes less essential for support since the many peripheral roots that have developed offer stability while maximizing access to nutrients. Tree taproots extend 10 to 20 feet deep, although in extreme cases they can grow much longer. The taproot in one fig tree in South Africa, for example,

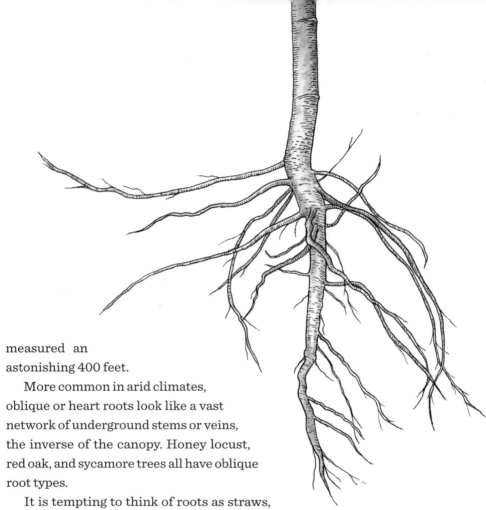

measured an astonishing 400 feet.

More common in arid climates, oblique or heart roots look like a vast network of underground stems or veins, the inverse of the canopy. Honey locust, red oak, and sycamore trees all have oblique root types.

It is tempting to think of roots as straws, but surrounding water isn't sucked up per se. Instead, water is absorbed through capillary action,

Tree roots

a process that allows liquid to flow up through narrow spaces, even against gravitational forces—in this case, up the trunks and into the tops of trees. Nothing is sucking on anything as with a straw in a glass of water. The surface tension of water helps the liquid travel upward into the highest branches.

Underground Web

The "wood wide web" is a concept that gained momentum with the publication of *The Hidden Life of Trees* by German forester Peter Wohlleben, but Dr. Suzanne Simard, a forest ecologist with the University of British Columbia, coined it. The term reinforces the idea that there is more going on beneath the surface, even in a forest of towering trees. This underground "web" of information is facilitated by a vast network of thin strands called mycelium. These threadlike strings, known collectively as the mycorrhizal network, are fungal organisms. They've formed mutualistic partnerships with the tree roots, and each participant in the arrangement gains something from the collaboration. Many people vastly oversimplified this concept, which went viral on social media.

Research has shown that trees shuttle water and nutrient resources between each other along this mycorrhizal network. Along the way, the fungi take 30 percent of the tree's photosynthesized sugars. Trees inoculated with these fungal associates receive more nutrients from the soil. Older trees have more mycorrhizal connections and are more likely to have an abundance of resources at their disposal than younger ones do. The ability to share these resources with young saplings that grow nearby is remarkable enough, but in some studies, it seems that trees shift nutrients at higher rates to support the saplings they are genetically closer to.

Dr. Simard, who has been working for decades to understand these exchanges, also studies how mycorrhizal networks transfer knowledge across species. It's not just Douglas-fir trees helping other Douglas-firs. Exchanges also occur between

Wood wide web

different trees. Simard's research uses stable isotope markers to track exchanges between Douglas-fir and paper birch. Her TED Talk on the topic, listed in Resources, has more than 5 million views.

Allelopathic Competition

Plants aren't defenseless from competition and herbivory. One mechanism for protection is allelopathy, or the production of biochemicals that are toxic to other organisms. These chemicals are unique to specific plant species, and their level can vary within individuals. Sensitivity to allelopathic compounds is also far from universal, so outcomes can range from decreased seed germination and stunted growth rates to full toxicity. If you find a bare patch of soil around the base of a tree, chances are allelopathic competition is playing out right under your feet.

Just as mowing the lawn doesn't kill your grass, herbivory (grazing and browsing) by animals isn't necessarily fatal for plants. It can even stimulate growth in many instances. Many parts of the plant (for example, nuts, fruits, drupes) are enticing to herbivores, which is one way that seeds get dispersed. Yet, discouraging consumption of certain parts of the tree is physically built into many tree segments. Spines and thorns

Black walnuts

adorn trunks and leaves of many species. Allelopathic defenses are usually hidden, but they are far from inconsequential. Tannins and alkaloids can also help protect a plant from being consumed.

Allelopathic compounds can limit competition between plants too. The black walnut is a well-known example. The genus, *Juglans*, hints at the tree's defenses. Black walnuts produce high levels of juglone—a toxic chemical most concentrated in the buds, nut hulls, and roots. Smaller amounts are present in the stems and leaves. The understory—often a combination of shrubs and saplings—is usually nonexistent beneath black walnuts as many of these plants are susceptible to juglone.

This technique for minimizing competition is also a hallmark of many invasive plant species. From tree of heaven (*Ailanthus altissima*) to garlic mustard (*Alliaria petiolate*), some of the most pervasive nonnatives rely on chemical suppression to maintain the upper hand. In addition to allelopathic actions, invasive species often grow very aggressively early in the season and produce high volumes of seeds. Cumulatively, these characteristics can lead to monoculture stands of plants that don't support native biodiversity.

Bald Knees

Baldcypress is an icon of the Southeast. It is the state tree of Louisiana, but the native range of this stately giant extends north to Delaware, up the Mississippi River to southern Indiana and Illinois, and west to Missouri. However, the hardy tree has been planted outside of this historic distribution. Despite its prevalence in backwater swamps, the plant can thrive in drier upland habitats. This adaptability is somewhat puzzling to scientists. You see, the knobby knees of baldcypress are far less common on trees that grow away from water.

Cypress knees are modified lateral roots. Instead of growing down into the subterranean, they stick up as thick projections. For a long time, researchers assumed cypress knees could increase roots' ability to transport air to the rest of the tree. These pneumatophore-type roots—lateral roots that grow upward and supply air to the plant—are found in mangroves and other plants that also inhabit saturated soils, without pockets of air. Closer investigation has since ruled out this hypothesis because the pneumatophoric tissues aren't present in cypress roots, nor are lenticles—the pores that allow for gas exchange in trees. Cypresses growing in deeper water lack knees, another adaptation that rules out the pneumatophore theory.

The lack of knees on trees growing in deeper water is also problematic for theories about how these growths function. If the knees add stability in wet soils, it would be reasonable for such a tree to have them no matter how deep the water is. A variation on this thought is that the extended roots help support the tree as it grows by trapping sediment. Younger trees don't have knees; cypresses may not start growing knees until they are at least a decade old.

While the function of cypress knees is perplexing, many other unique characteristics are better understood. Cypresses are one of the few deciduous conifers, even within the Cupressaceae family, which includes junipers, cedars, redwoods, and sequoias. The thin, feathery green needles of the plant turn a rich cinnamon brown in the fall before they are shed for winter. (One variety of baldcypress, sometimes called pond cypress, has scaled leaves like junipers and cedars.) The trees are bald in the winter, but the name applies

Baldcypress knees

throughout the year. Male flowers in the form of long, dangling catkins are conspicuous before the needlelike leaves fully regrow in the spring.

Its swampy habitats spared cypress forests from the earliest days of logging, but by the early 1900s, the mills turned to the trees and their knees. Now, it is nearly impossible to find old-growth individuals or untouched domes and strands. Cypress has been heavily harvested for everything from shingles to mulch. Louisiana has lost around half its cypress swamp, and the numbers in surrounding states aren't any better. Cypress is especially vulnerable to high levels of harvesting because the species is slow to grow and has a relatively low rate of natural reproduction.

Mangrove Fingers

The long fingers of dangling mangroves conjure thoughts of the tropics. Three species of mangrove reach the northern limits of their ranges along the Gulf Coast in the southeastern US. Everglades National Park in Florida hosts red, black, and white mangrove swamps.

The red mangrove is the most widespread in the region. This species represents the transition from water to land with lanky roots that look like stilts protruding into the substrate beneath the tidal waters. These specialized prop roots, roots that grow aboveground, allow the plants to survive in areas that are inundated with water and experience the tidal cycles. The roots supply air to plants that grow in anaerobic soils that don't have pockets of air. The branchlike roots also help anchor and support the trees.

While red mangroves can reach heights of 80 feet under certain growing conditions, they are much more adapted to grow outwardly,

Prop roots of a mangrove

and trees in the Everglades rarely grow taller than 20 feet. Instead, these "walking trees" send out horizontal prop roots, slowly shifting their position. The resulting tangle of roots offers shelter and protection for countless aquatic species from invertebrates to fish. The extensions also trap sediment, building coastal soils as it accumulates. Wind pollinates red mangroves; the seeds then germinate on the fertilized trees. These elongated propagules—or new plants—eventually drop off and either plop into the mud or drift away on water currents where they settle and grow elsewhere.

Dark- and scaly-barked black mangroves thrive in a narrow zone upland from the red mangroves. To combat saturated wet soils, black mangroves have roots that poke up above the ground like snorkels. These pneumatophores, or air-breathing roots, aid the plant in surviving in this saltwater-rich environment. Black mangroves can also remove salt from the water; they deposit the salt crystals on their leaves. Black mangrove is the most widespread species along the Gulf of Mexico.

The least cold-tolerant of the northern mangroves, the white mangrove grows farthest inland. It is also the species that looks most like a traditional tree with an underground root system. Its leaves set it apart. The base of each oval leaf has two nectaries—little nubs that secrete a sugary substance that attracts pollinators such as butterflies. These nutrient-rich treats lure ants, which in turn, benefit the white mangrove by protecting the tree from certain pest species.

Pando Grove

The underground connections of aspen trees spread far and wide. Aspen grows in relatively uniform, even stands, and these clumps are distinct. Sometimes, though, mixed stands that include a few conifer trees break up the monoculture. Even though a few aspen stands may include multiple individuals, an entire stand is usually a single "tree"—the separate trunks are often all cloned shoots extending from a shared root system.

The largest grove of quaking aspen, *Populus tremuloides*, called Pando (Latin for "I spread"), is a single tree covering more than 100 acres in central Utah's Fishlake National Forest. Based on DNA evidence, Pando is the largest living organism. There are more than fifty thousand individual trunks, of which the oldest stems are roughly 100 to 150 years old. Pando's root system, however, has been alive for *much* longer—it dates back eighty thousand years. Despite the tree's longevity, Pando's health is declining overall.

Tree trunks in the stand are not regenerating because of multiple factors. Pando shoots are heavily grazed, and this lack of recruitment is hampering the stand's ability to maintain itself. The grove is also susceptible to fungal disease and insect damage. Land managers have also been observing encroachment by junipers. And shifting fire regimes could be one more factor. Small, controlled burns and fencing have shown some success in helping Pando regenerate, but the ancient being might simply be dying slowly of old age.

Pando's decline isn't an anomaly. Many aspen stands across the West show a similar lack of regeneration. This slow decline is being accentuated by the rapid loss of stands in some areas, including in Colorado. First recorded by land managers in 2007, Sudden Aspen

Decline (SAD) affected 13 percent of the aspen in the Centennial State in just a few short years. Drought conditions could accelerate the dieback of aspen, as they are left without enough resources to fend off pathogens.

Quaking aspens

State Trees

You can probably rattle off the name of your state's official bird, but do you know its official tree? All fifty states and the District of Columbia have declared a symbolic timber species. The trendsetter for honoring trees was Texas. Although the concept of adopting a state tree had been around for decades, the Lone Star State declared the pecan the state tree more than a century ago in 1916.

The process of officially recognizing a tree in New York played out as quite a prolonged saga. Students voted in 1889 to honor the sugar maple, but it took decades for the state government to consider honoring a tree this way. Despite the vote of the kids, and perhaps inspired by designations of oak trees in Maryland in 1941 and Connecticut in 1947, New York legislators proposed the red oak as the official state tree; however, that proposal failed. In 1956, the New York governor pushed to recognize the sugar maple, an idea that didn't sit well with officials in other syrup-producing states from Maine to Minnesota. While Vermont and Wisconsin had, in 1949, both previously adopted the sugar maple as their official tree, several states were also vying for syrup-bragging rights. A maple taste-off was held in New York; nine states and a Canadian province were represented. The contest ended in a tie between Vermont and Michigan with the home state—New York—taking third. Shortly after the competition, New York certified the original wishes of those long-ago schoolchildren and adopted the sugar maple as the official state tree on April 10, 1956.

Trees were also chosen by students in popularity contests in West Virginia, Montana, and Utah, among others. West Virginia and Montana assigned state trees in 1949: students in West Virginia picked the sugar maple, while Big Sky kids chose

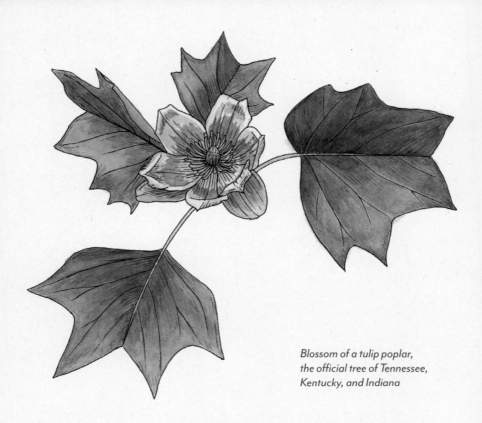

Blossom of a tulip poplar,
the official tree of Tennessee,
Kentucky, and Indiana

the ponderosa pine in 1908, but it took the government of Montana more than forty years to make it official. Utah adopted the blue spruce in 1933, but this stately conifer was dethroned in a vote in 2014. The grassroots—*er*, tree roots—effort was student led. They felt that because the aspen tree is found in all twenty-nine counties in the state, it was a more representative choice.

Nevada double dips on tree designations. Both the single-leaf piñon pine and the bristlecone pine are officially recognized. Tennessee also has multiple honorees. Rather than toppling the poplar after fifty-five years, it added an official evergreen to the ranks of state tree. So, the state has both a state tree (tulip poplar) and a state evergreen tree (eastern red cedar). Tennessee shares the tulip poplar as its official tree symbol with Kentucky and Indiana.

Banyan Tree of Lahaina

It can sometimes be difficult to see the trees through the forest, although in many instances, single trees become iconic. According to Hawaii's Lahaina Restoration Foundation, Queen Keōpūolani, the sacred wife and widow of King Kamehameha, initiated a Protestant mission in Lahaina in 1823. As a gift honoring the mission's fiftieth anniversary, Sheriff William Owen Smith planted an 8-foot-tall banyan tree from India at this location in 1873. Fast-forward a century and a half, and this banyan tree is now a goliath, dominating the courthouse square in downtown Lahaina.

On the west coast of Maui, the Lahaina banyan is a focal point. The tree now stands more than 60 feet high, and more impressively, it extends outward for an entire city block. Banyan trees, sometimes called banyan figs or strangler figs, can start out as epiphytes, which are

The goliath banyan tree in Lahaina

plants that grow on other plants but are not parasitic (see Epiphytes in Branches & the Canopy). The plant can extend aerial roots from branches down into the ground. These roots can then divide into separate plants, or expand the reach of the original trunk. The long, waving branches and extensions of rooting trunks make it remarkable from every angle. The Lahaina banyan now has forty-six distinct trunks. More than a football field's worth of canopy provides shade for many events. It is the largest representative of the species in the United States. To walk around the Lahaina banyan is like walking around a standard outdoor track; one lap is about a quarter of a mile.

Starting in 1802 and lasting forty-three years, Lahaina served as the capital of the Hawaiian Islands, per a declaration from King Kamehameha. The capital has since been relocated to Honolulu, but the banyan tree of Lahaina continues to thrive.

Rot

Underground vegetation is an essential link to nutrients in the soil, but roots face many pathogens and disease vectors. Root rot attacks the underground levels of trees, while butt rot targets the base of the tree. They can be caused by the same factors, and together they drive stunted growth and mortality in forests. Rots can cause trees to topple over easily. In residential settings, this can be problematic, but in a more natural setting, such events are often just part of the life cycle of the forest.

Common ailments responsible for rot are the presence of numerous fungal species. The fibrous growth of these fungi in the soil extends outward in a way that resembles the expansion of fairy ring mushrooms: a fungus grows underground, and then it sprouts small threads (mycelium) that spread away from the center. Aboveground, there is no evidence of the

Shelf fungi

original fungus, but a ring of mushrooms appears when conditions are right.

Pockets of dead or dying trees can reveal rot. Conks or bracket and shelf mushrooms are other indicators of a potential outbreak of rot. Many of these trunks develop snaps along their lower portions. Rot-infested trees tumble at random, creating a pick-up-sticks tangle, whereas trees damaged by blowdowns all topple in the same direction.

Even when the rot doesn't kill the individual, these weakened plants are easy targets for a secondary attack. Bark beetles and other insects take advantage of these vulnerable trees. The plots of sick trees create a mosaic on the landscape and reset the successional process (see Succession in Snags & Coarse Woody Debris).

Water infiltrates through the root system

Water Transpiration

Trees can act like gigantic sponges. The canopy intercepts precipitation as it falls from the sky. Root systems slow infiltration rates and directly take up moisture. A single mature specimen can consume hundreds of gallons per day. Research conducted in Minneapolis has demonstrated that, on average, each tree diverted 1,685 gallons of stormwater per year.

The amount of available water can limit tree growth. Additionally, water consumption by trees isn't overly efficient. A tree is constantly balancing its intake of carbon dioxide with its output of water. Transpiration rates hover between roughly 5 and 10 percent of all the water they take in each day. A small amount of water is used for plant growth, while much of it plays a role in keeping the tree at an appropriate temperature. Leaves lose plenty more water via evaporation.

It's easy to think of a pipeline running from the roots to the leaves like a straw, but that's not quite how it works. The xylem cells, which act mainly to transport water from roots to the rest of the plant, aren't actively absorbing water. Instead, negative pressure facilitates the flow. As water evaporates out of the leaf, the moisture from adjacent cells is drawn up. This chain of events plays out all the way down to the tips of the roots, thanks to the linked nature of the inner tree cells affecting nearby cells. So, warmer air temperatures at the top of the plant drive increased nutrient flow inside the tree, thus creating a suction effect.

In areas with extended winters dominated by cold temperatures and fewer hours of daylight, many trees go nearly dormant, or inactive, for this part of the year. To prepare for the change of the seasons, the plants develop antifreeze-like substances. Also, trees remove most of the sap from their trunks, which helps prevent freezing, expanding, and potentially rupturing as temperatures drop. Frost cracks still do occur occasionally.

Longleaf Grass

Once an important source of lumber in the Southeast, the longleaf pine remains the state tree of Alabama, although it is now an ecosystem in peril. The species is down to less than 3 percent of its original range, which was once a vast 90-million-acre swath stretching from southern Virginia to Florida and west to Louisiana and neighboring Texas. Small patches are all that remain of this critical habitat. Since the wood was prized for building ships and railroads, the forests were heavily harvested. By the early twentieth century, few longleaf pines were left. The species has been especially slow to recover partly because foresters planted other species in its place. Its slow-growing nature means it is a denser wood. It's straighter, stronger, and thicker—all characteristics that make it desirable for lumber products. However, slow-growing wood offers less of a short-term economic benefit compared to faster-growing species.

The life history of the longleaf pine also works against it in restoration efforts. As a fire-dependent tree species—meaning fire helps the cones open and release seeds—opportunities for regeneration were limited for much of the 1900s. In 1935, the US Forest Service established the "10 a.m. policy," which meant the goal of firefighting

Longleaf pine

was to suppress any forest fire, even small ones, by 10 a.m. the morning after it started. This policy prevented many fire-dependent plants from fully rejuvenating. Longleaf pine nut seeds germinate best on exposed soil, and ideal conditions follow a wildfire that occurs naturally. As forest duff (partly decayed debris on the forest floor) accumulates between disturbances, longleaf seeds probably won't sprout. Unlike most pines, when the seedlings do sprout, longleaf pines don't resemble miniature trees initially. Instead, they go through a fire-resistant grass phase for the first few years. At this point, the "longleaf grass" trees are using resources to develop a strong root system before a trunk starts to grow.

Longleaf pines are hardly recognizable as trees during this initial phase, but the towering giants will reach great heights over the course of their three-hundred-year lifespan. Mature trees stand nearly 100 feet tall and can easily reach 3 feet in diameter. Longleaf pines famously grow straight. Red-cockaded woodpeckers and indigo snakes are the poster critters for the forest, but stands of longleaf pine support more than thirty other threatened or endangered species.

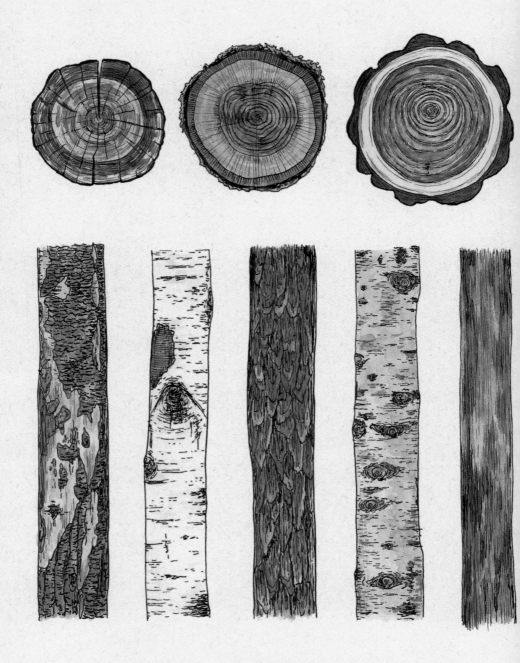

Trunks
& Rings

In this section, we'll discuss the inner workings of trees and differences between the wood of various species. The main beam structure of a tree is its trunk. This wooden stem is one characteristic that separates trees from other plants. The trunk is also where the nutrients and water are transported to the rest of a tree's growing parts, even though much of the trunk is dead tissue.

Tree rings are a recurring theme in this section. We can learn a lot by examining rings—not only about that tree's anatomy but also about climate shifts and conditions from decades and centuries past. Rings help botanists reconstruct the past.

Bark can be a diagnostic tool for identifying some species and serves as the skin of the tree. It is not impermeable, however, making trees susceptible to diseases, infections, and infestations. One example of a tree species succumbing to disease is the American chestnut. Once an icon with an extensive range between the Atlantic Ocean and the Mississippi River, because of a blight ailment, it is now functionally extinct. This sad tale is summarized in this section, as are the efforts to save it.

Finally, two very different forests are the focus of a pair of entries. The vast boreal forest and stands of doghair, or young, forest—both are important habitats for many specialized critters.

Microscopic Tree Cookies

If you've ever looked at a cross-section of a tree (smaller ones are sometimes called tree cookies, larger ones, wedding cake display stands), you've probably noticed tree rings. You can learn a lot about the life of a tree from these slabs (see Seeds of Knowledge: Dendrochronology, later in this section). But if you put a much thinner slice under a microscope, a tree's inner workings can be revealed.

Not all wood looks the same. Individual cells cluster together, and different types of cells serve various functions. The cells of the inner core are no longer living, and the outer bark tissues are also not alive. The active living cells are within the cambium tissues, just underneath the bark, where the phloem and xylem are generated. Phloem is the tree's route for transporting food. Glucose sugars produced by photosynthesis in the leaves travel along the phloem and are distributed throughout the branches, trunk, and roots. The xylem is the water canal. These cells bring water and nutrients up from the roots all the way to the

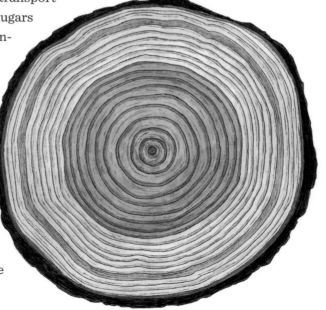

Cross-section of a trunk

leaves in the upper canopy. Sap is the flowing, viscous river of nutrients inside the tree and comprises both the phloem and xylem products. The composition of these materials changes with the seasons.

The vascular tissues (the cells that conduct the tree's fluids) help support the plant and aid in this nutrient flow, even when working against gravity to get water found underground to the upper treetops. It is easy to imagine the sap moving inside a tree like a milkshake getting slurped through a straw. Instead, the liquid moves slowly through porous cell membranes with capillary action (see Root Systems in Roots, Buttresses & Knees). Imagine what it would be like to spill your morning coffee, and then toss a folded towel on the puddle. The many layers of fabric, not just the bottom sheet that is in direct contact with the spill, would absorb the liquid.

Growing Up & Out

Even if you aren't measuring in board feet, tree growth is impressive. This seemingly solid material expands and elongates. Increased height and extended branching happen at meristem points (or growing tips). These meristematic tissues are where cell division occurs.

Most of a tree is dead wood. Old xylem dries out, dies off, and becomes part of the heartwood core, while the bark incorporates phloem from past years. The reintegration of xylem and phloem contribute to a trunk's expanding girth and the thickening of branches.

The active living tissues in these regions of the tree are part of the lateral meristem.

Terminal (or apical) meristems are located at the tree tips, on both the roots end and the shoots end. These cells are capable of growth and division. Both the roots and the canopy branches lengthen each year. By inspecting the twigs, you may even see the individual segments of growth from years gone by. These segments aren't as reliable or as easy to see as tree rings, but they can be obvious, especially for the most recent year or two of growing seasons. Knobby nubs often delineate the sections for each year. Maximum tree height is determined by genetics, but local growing conditions, such as late spring snow or dryer-than-normal summers, help dictate an individual's ultimate reality. Not all trees will reach their genetic potential, and it's not their fault—they likely didn't have access to enough water or sunlight.

Natural processes can cause these growth parameters. Branches fall with some regularity, and especially stiff winds can snap the top of a tree right off. Tree growth response is especially evident in windswept coastal

Tip of a tree branch

areas and mountaintop forests (see Krummholz in Branches & the Canopy). Arborists embrace these growth factors to prune trees methodically. An extreme version of this is bonsai trimmings, but other applications include altering growth away from homes or powerlines. Similarly, low-hanging branches can be cleared out to open up the understory. The upper meristems produce hormones that regulate the other branches, so damage to this section can slow growth, resulting in bushier trees. This can happen naturally, or arborists can manipulate the growth by pruning.

Hardwoods vs. Softwoods

It is perplexing that so many people categorize trees into the broad categories of either hardwood or softwood. Botanically, the hardwoods include the angiosperms or broadleaf trees, such as oaks, maples, and hickories. Softwood species are the gymnosperms or conifers, including pines, spruces, and firs. Admittedly, these terms are applied in woodworking more than in forestry.

Hardwood species grow slower, resulting in denser wood. Hickory is one of the sturdiest hardwoods in North America, but there are exceptions. Balsa is a hardwood, yet it is one of the lightest woods around. Other species are hard softwoods. Southern pines including longleaf, slash, shortleaf, loblolly, and yew are all hard softwoods.

Hickory tree

Dendrochronology

If you've ever counted the rings of a tree, consider yourself an amateur dendrochronologist. Roughly from the root words *dendro* (tree), *chronology* (time), and *ologist* (researcher), dendrochronologists do more than merely tally up lines in the wood. Instead, these scientists can read the story of the trees and climate by examining details and unlocking the codes left behind within the circles.

Seasonal growing rates determine the growth bands for an individual tree. Since growth is consistent throughout the year in the tropics, for example, tree rings don't register in these locations. But where seasons vary, earlywood cells are laid

Close-up of tree rings

down at the start of the spring growing season. Summer is the season of growth. In favorable years with enough moisture, typical temperatures, and sufficient sunlight, more cells develop, creating thicker rings. Drought conditions and other stressors can limit wood growth in any given year. Thicker-walled cells are created as the growing season wraps up. They often appear as darker bands and are the easiest to count. Together each pair of light- and dark-growth bands represents an annual growth ring or a year of the tree's life.

The fluctuation in the widths of tree rings is one piece of the climate puzzle that dendrochronologists rely on as they unravel mysteries from hundreds or even thousands of years ago. The drama of life and death plays out in each tree too. Fire scars that damage a tree without killing it are recorded in the wood. Timbers harvested for construction projects such as buildings and boats can be dated with precision, combining the fields of dendrochronology and archaeology. Researchers also look to make future predictions with the data patterns they've collected. A team of University of Arizona ecologists, for example, predicts changes in climate will decrease the size of ponderosa pine trees in the state in the coming decades, which has ramifications for carbon retention.

One of the densest woods is ironwood. This title conjures the species *Parrottia persica,* which is native to Iran, although there are at least five other species of trees (in five different families) that go by the same common name in North America. Tropical hardwoods top the scales for density.

In botany, *resistance* refers to pests and disease, but in carpentry, the Janka Wood Hardness Scale, a method for ranking the hardness of different types of wood, quantifies the resistance of wood to wear and denting. Various factors alter the strength of wood, including the direction of the grain.

Another place where the differences between hardwoods and softwoods are evident is in a campfire or woodstove. Conifers and other softwood logs burn quickly and emit more resins that, if allowed to accumulate, can make a home susceptible to chimney fire. In most instances, people prefer hardwood logs because they burn hotter, cleaner, and slower.

Old, Gnarled Sentinels

Trees are notably long lived, but the oldest of the elders are the bristlecone pines. Some individual bristlecones are around five thousand years old. For Casey Clapp, co-host of *Completely Arbortrary*, the podcast dedicated to tree appreciation, "It's the single greatest tree there's ever been." He continues, "I'm just such a fan of history, and this thing has watched it all."

Great Basin or intermountain, Colorado or Rocky Mountain bristlecones, and the foxtail pine are nearly identical. Known collectively as the bristlecone pine complex, these species inhabit separate mountain ranges from California to New Mexico. They all eke out their existences on the edge of survivable extremes. These trees grow at the upper reaches of the timberline in twisted krummholz-like postures. Limited nutrients and short growing seasons translate to slow growth. The size of these gnarled trees does not reflect their age.

Bristlecone pine

For them, anything above 30 feet tall is a towering giant. They may put on only an inch of girth every century.

The *Pinus longaeva* is the oldest of the cohort. Many individuals are at least four thousand years old. The current candidates for the oldest trees are Methuselah—a specimen believed to be approximately 4,850 years old, whose location is undisclosed to protect the tree from visitors and vandals—and a neighboring tree dated at a couple hundred years older. Inyo National Forest and Great Basin National Park have protected stands of bristlecones, yet the species as a whole is struggling. They've survived for centuries, but prolonged drought and bark beetle infestations are threatening to wipe the species out.

In 1964, an individual tree was chopped down in the name of science, causing an uproar still widely talked about, although it didn't change the species' prospect of survival on a population scale. A field botanist was conducting dendrochronology research by drilling out sample wood cores. Usually, it's a straightforward process to core out the straw-sized samples of wood, which doesn't usually harm the tree, but the dense wood of the bristlecones causes the tool to jam up. A second tool also got wedged in the tree. To retrieve the tools and collect the required wood samples, the botanist then obtained permits and permissions from the Forest Service to ultimately cut down the tree. When the researcher counted the tree rings, he discovered that this now dead tree would have been the oldest living thing documented at 4,862 years old. The tree, known as Prometheus, lives on as a legend, even though its life was cut short.

Tree-skeptic and *Completely Arbortrary* co-host Alex Crowson thinks the bristlecone pine sounds like an incredible and beautiful looking tree, but he reserves judgment until he can experience one firsthand. "I need to place my hand on it and feel the ghosts of the ancients," he says. "They are totally a haunted-looking tree."

"Wooden" Trunks

A tree's wooden stem is a defining characteristic. It is also the reason palms aren't trees. In the Arecaceae family, palms are botanically monocots, which means that only one leaf comes out of the germinating seed. The structures of the trunk in plams differ from that of classic trees. The pseudo bark is made up of a thin layer of epidermis and a section of sclerified tissue (having developed thicker, more rigid cell walls) called the cortex. A cross-section of a palm tree shows no growth rings; instead, it looks uniform with a somewhat spongelike appearance. Within this matrix of cells are the vascular bundles that include the xylem and phloem.

While the inner heartwood of traditional trees is made of dead cells, the central cylinder of the palms is living tissue. Their fibrous nature makes them well suited for the high winds of hurricanes. The plants sway farther than typical trees and are less likely to snap off,

Sabal palm

but they are still prone to uprooting and toppling over. Palms don't grow out each year; instead, they grow up. A distinctive feature on the palms is that the leafy structures—the fronds—grow out of the very top of the plant. The scaled appearance along the stem is where leaves were once attached when the palm was younger.

Despite not being a tree botanically, the sabal palm is the state tree of South Carolina and Florida. Another confusing example of a tree that isn't a tree, despite its name, is the Joshua tree. These iconic plants are related to yuccas. Similarly, bamboo is a plant that has a woodlike stem, yet it is in the grass family.

Sapsuckers

Sapsuckers are members of the woodpecker family, so it's no surprise they have chiseled beaks. They wield these tools differently than other woodpeckers though. As their name implies, in addition to the typical insect fare woodpeckers eat, the sapsuckers are after the juice of the tree; they carve out uniform rows of small sap wells in the bark. Woodpeckers can stick out their tongues—a rare talent in the world of birds. Sapsucker tongue tips are like tiny paintbrushes. With each flick, the birds load the sap onto microscopic bristles. They also eat the thin cambium zone of living tissue under the bark.

Sapsuckers are predominately black and white, with varying amounts of red on the head and shades of yellow on the body. Long

Sapsucker making sap wells

white stripes are visible on the folded wings of perched birds. This feature helps distinguish them from the members of the woodpecker family, such as downy woodpeckers and hairy woodpeckers, that people are often more familiar with.

There are four species of sapsucker in North America. Until 1983, red-naped, red-breasted, and yellow-bellied were all considered a single species. While most woodpeckers stick to a local area throughout the year, sapsuckers are migratory. Yellow-bellied sapsuckers breed in the northern forests and winter in the southeastern United States and northeastern Mexico. Watch for this bird during migration anywhere east of the Great Plains. Nesting from southern British Columbia to central Arizona and New Mexico, red-naped and Williamson's are sapsuckers of the western mountains. The red-breasted sapsucker ranges along the Pacific Coast states and provinces from Southeast Alaska to Southern California. They shift on the landscape from summer to winter but remain in the northern reaches of their range year-round.

The sap wells made by sapsuckers are magnets for other species. Numerous species of hummingbirds have been documented visiting the seeps to partake in the sap and snag insects that have also been drawn in. Rufous hummingbirds sometimes nest near red-breasted sapsucker trees. In the east, the ruby-throated hummingbird's spring migration coincides with the sap flows created by yellow-bellied sapsuckers. Birches and maples are favorites of yellow-bellied sapsuckers, but the species has tapped more than one thousand species of woody plants. Williamson's sapsuckers seem to use only conifer species for their buffet lines.

Hollow Homes

The inner cells of a tree are no longer living. Natural decay means that portions of the stem in healthy trees may be hollow. These tree cavities aren't just for feathered inhabitants (see Cavities in Snags & Coarse Woody Debris); instead, these hollow trunks are prime habitats for many mammals as well.

Porcupine on a branch

Temperatures within hollows can be wildly different from air temperatures. This buffer from temperature swings is noticeable in all seasons, as cavities stay warmer in winter and cooler in summer. Cavities also offset temperature fluctuations between night and day.

Not all squirrels are arboreal, but the ones that are take advantage of hollow homes. In addition to these cavernous homes, they make dreys, or temporary shelters of woven twigs and leaves. Young squirrels born in dreys are less likely to survive than those growing up inside the protected hollows. Similarly, raccoon kits staring out from a hole in a tree is a classic nature scene. Baby raccoons remain in this natal den for seven weeks on average. After that, the family moves every few days, finding new locations to spend the night.

Never the most active of critters, porcupines hunker down inside a tree for long stretches at a time. Piles of scat accumulated at the base of a tree hollow can reveal porcupine haunts. Look for compressed, sawdust-filled, cylindrical pellets.

A surprising treehouse dweller is the gray fox. This is the only North American canid species that regularly climbs trees. Semiretractable claws help them scamper up and down tree trunks.

Some species of birds and bees are drawn to hollow cavities. Many bats also require forest habitat. Some species prefer to roost inside the trees and snags, while others tuck in beneath the bark, and a few rest in the leafy canopy.

Smooth Birch Bark

Young trees often have smooth bark, even on species that ultimately develop a textured trunk. Some species, such as beeches and aspens, are characterized by this smooth bark for their entire life cycle. One of the most familiar smooth-barked trees is paper birch. Smooth, white bark is a unifying feature between the unrelated birch and aspen, but the bark of paper birch can peel and flake, while aspen bark remains tight and intact. As tempting as it is to give birch bark a tug, it's best not to disturb this fragile tissue.

Birch bark may have a tinge of green. Along the trunk and in the leaves, birch, as with many species, can conduct a low level of photosynthesis. While photosynthesis in tree bark is only about 30 to 50 percent as effective as photosynthesis carried out in the leaves, in spring, before leaves emerge for the season, it can be especially valuable for an individual tree.

You'll notice in birches that the bark isn't entirely uniform

Peeling bark on a paper birch

because black scabs contrast with the white bark. Aspen markings are larger, more sporadic, and appear knotty and knobby. In birch, look for small, uniform, horizontal black markings, which are lenticels that aid in gas exchange. (Other species also have these lenticels, especially on younger trees, but they are more difficult to see.)

The birch family includes around thirty-five species of cold-tolerant plants in the northern hemisphere. The name "birch" stems from the Germanic language. Following a disturbance, birches can be some of the earliest woody plants to reestablish.

Paper birch is the provincial tree of Saskatchewan in Canada. Either sap or chunks of bark can be transformed into birch beer or syrup, tree products notable for having a wintergreen essence.

Forest Products

Forests and trees have intrinsic value, including providing services essential to the functioning of ecosystems and serving as important habitat for countless critters. These "ecosystem services" are the life-sustaining benefits that nature provides. But as a society, we use trees as a commodity in many, sometimes surprising, applications. For example, did you know you can find tree products in disposable diapers?

Wood is the most obvious example of a product derived from the lumber that was once a tree. As you can imagine, the spectrum of woods matches the diversity of trees. The distinction between hardwoods and softwoods (see Hardwoods vs. Softwoods earlier in this section) is a broad categorization, but to a master carpenter, each type of tree and every single board that comes from it has a unique character. In some instances, perfectly symmetrical grains are desirable, but gnarls that give wood character can often be incorporated into projects and designs. Burl bowls and wooden spoons are certainly making a comeback in the DIY and craft scenes. Heart rot can provide a natural stain and accent the coloration of the lumber.

Wood used to create musical instruments sings a different tune. Also known as tonewood, instruments' components are carved and shaped into one-of-a-kind pieces of functional art. High-end instruments are carved out of top-dollar trees. For legendary Stradivarius violins, for example, the sound quality can be attributed to many factors. These centuries-old instruments are unique, thanks to the wood—always maple on the back and spruce on the front—the processing chemicals, and the countless vibrations played on them. In guitar circles,

Stradivarius violin

instruments made from "The Tree," a mahogany from Belize, are things of beauty, and for roughly $30,000 to $40,000 you too, can own one. Slash from Guns N' Roses has one topped with a panel made from a three-thousand-year-old Sitka spruce.

The more processed the lumber, the less it resembles a tree. Sheets of plywood and particleboard are reengineered wood, in which veneers and chips are compressed together to create "boards." Breaking down the fibers even further gives manufacturers wood pulp—the basis for papermaking. In the United States, 85 percent of the paper produced comes from species of conifer. These trees have longer fibers and create stronger paper materials. A single tree can create more than eight thousand sheets of paper. Approximately four thousand reams of copy paper can be grown annually on a 100-acre forest. Fluff pulp is hardly recognizable as the trees it once was— spruces and pines—but this highly absorbent material is used in specialized instances, including diapers.

Tree Treat

One of the sweetest of the tree treats is certainly syrup. While maple is the most famous flavor of this delicious delectable, sap harvested from a number of tree species, including beech, birch, sycamore, and boxelder, can be boiled down into syrup.

Tree tapping is a late-winter tradition across much of New England, the Upper Midwest, and Canada. The ideal temperatures

Tapping maples for syrup

for sap flow are above freezing during the day, but below freezing at night. All winter long, tree sap has been in the root storage units. At this earliest phase of spring, the sap begins flowing upward toward the crown.

Some commercial-scale operations use miles of tubing and vacuum suctioning to maximize harvests, but for backyard sugaring, the tapping process is quite simple. Drill a small hole approximately 2 inches into a suitable tree, and then gently tap a hollow spile (rhymes with "smile") into place. The sap will drip out of the spile and into a collection bucket or bag. Daily outputs can be a couple of gallons of sap per tap, but that amount would yield very little syrup. For sugar maple trees, a reliable estimate is 40 gallons of sap per gallon of syrup. Making syrup from other species could require nearly double that amount of tree juice. Yet, given that the annual tapping season has been extending for a few weeks as the climate shifts, even a hobbyist can gather enough sap for a modest syrup supply.

To concentrate the sugars and flavors, the excess water content is boiled off as steam. Sugar shacks are the traditional processing sites, but for a backyard syruper, a propane burner can get the job done. Broad pans work best because they have a larger surface area for evaporation. The rolling boil of sap produces vast amounts of sticky steam, so syrup making is an outdoor activity. Figure about an hour of boiling for every gallon of sap. It's not difficult work, but it is easy to scorch the sap at the very end. Many folks move inside to finish the syrup on a more sensitive stove burner. A candy thermometer will help you hit the target temperature of 219 degrees Fahrenheit.

Maple syrup isn't just for pancakes and waffles. Try the treat as a glaze, drizzle it atop an ice cream sundae, or pair it with bourbon and sip it as a cocktail. Any way you enjoy it, it'll definitely taste better than any knockoff sugary, fake maple table syrup.

Chestnut Blight

Generations of Americans have missed out on a true icon of the eastern deciduous forests: the American chestnut. In the 1800s, this species accounted for up to 25 percent of the forest in hardwood ecosystems between the Atlantic Ocean and Mississippi River. While American chestnuts are still found on the landscape, the trees now exist primarily as an understory plant with shrubby growth patterns—thanks to a nonnative fungal disease—instead of being the stately keystone species it once was.

Chestnut blight developed alongside Eurasian chestnut species. When foresters in New York discovered it in the early 1900s, they were rightly concerned. The disease was likely brought to North America from Japan, hitching along in shipments of ornamental Japanese chestnut trees that evolved with the blight and so can fend off its maladies. In those chestnuts, a canker swells along the trunk, walling off the infection from the rest of the tree. The trees of the eastern United States and southern Canada, however, aren't equipped to deal with this blight. When American chestnuts are infected, the disease spreads quickly within the tree with erupting cankers that allow spores to be dispersed on the wind, putting neighboring trees at risk as well. Researchers estimate nearly 4 billion American chestnuts were infected from the 1880s to the 1930s. The root systems can survive, but the blight knocks the shoots back before they can grow to maturity.

Researchers are tackling this problem with efforts unfathomable nearly 150 years ago when chestnut blight first arrived in North America. One potential solution is to genetically hybridize

American chestnut trees with Chinese chestnuts to create blight-resistant plants. A second solution involves sophisticated transgenics. Through gene editing, researchers can use genes from wheat plants, which regulate an enzyme that can enhance a tolerance to blight, to share this resistance with chestnut trees. Biocontrol options, which use living pest control mechanisms, are also being explored to disrupt blight infestations. Together these techniques may offer ways to restore American chestnuts and return these forgotten giants to the eastern hardwood forests.

American chestnut

Boreal Forests

Despite accounting for nearly one-third of the world's woods, the boreal forest is unfamiliar to many people. These dense, dark woods of the north are sometimes referred to as taiga. The boreal biome is circumpolar, meaning it is found across the northern latitudes of North America, Europe, and Asia. The woods of Canada, Sweden, and Russia have a similar feel and, in some cases, even share the same species.

Conifer trees dominate the cold, subarctic taiga. Mixed stands of evergreen spruce, fir, and pine stretch for miles. Pockets of larch, a deciduous conifer, shed needles annually, while aspen, birch, and poplar provide sprinkles of leafy foliage. In many locations, a layer of subsurface rock and permafrost creates boggy muskeg conditions. Trees in the boreal forests have many cold-tolerant adaptations. Waxy needles help prevent them from drying out. Like an A-frame roof, the classic holiday tree shape sheds snow rather than accumulating it at the risk of breaking off branches.

The taiga is teeming with wildlife too. Roughly half of North American bird species rely on the northern woods in their life cycle. Boreal chickadees and boreal owls are both named for the region (which stems from Boreas, the Greek god of the north wind). Within the woods, Canadian and European lynx populations mirror the cycles of hares they prey upon. The largest members of the deer family—North American moose and the European elk—are the same critter, *Alces alces*. These black-bodied beauties are increasingly rare in parts of their range, as warmer temperatures alter the habitats they thrive in. Ticks can survive increasingly mild winters, and the

Wooden Words

Trees and wood appear in many familiar phrases. Even if you aren't a tree hugger, you're likely familiar with at least a few of these wooden words. How often have you heard about not being able to see the forest for the trees, for example? If you're ever in a sticky situation, you must try to get out of the woods. This phrase dates to at least 1792.

People can set down roots or be rooted in place. When things aren't going well, you might be barking up the wrong tree. When similar patterns play out across generations, the apple doesn't fall far from the tree. A version of this saying was around in Germany at least as far back as the 1500s.

People can visit someone else's neck of the woods or invite guests to their own neck of the woods. Sawing logs or sawing wood means to snore loudly, although nobody has ever been accused of snoring quietly. "Cutting the deadwood out" means getting rid of something that isn't working, but at least regionally, this adage appears to be fading from the lexicon, replaced by the similar phrase "cutting of dead weight."

"Setting the woods on fire" means something is a big deal, but this idiom is applied equally to "not setting the woods on fire," as in something being a letdown. A classic expression involves a defecating bear; however, the woods aren't really the subject of this phrase but merely the implied location of the deed. The first instance of this making it into print is in the sports pages of a newspaper. Players generally referenced bears obviously living in the woods in the late 1950s and early 1960s, but by 1966, the bears were either sleeping or crapping in the woods.

The ska punk band the Mighty Mighty Bosstones may not have ever had to knock on wood, but for the rest of us, this good luck ritual can be standard operating procedure. The tradition likely dates to at least—or perhaps even predates—ancient Indo-Europeans. According to some folklore beliefs, spirits lived within the various trees. Knocking on or touching trees initiated blessings and protections from the spirits. Although there isn't much empirical data to support the efficacy of such efforts, mind over matter may be just the boost you need. If the simple act of knocking improves your outlook, perhaps it can make you feel better.

What about the phrase "Don't worry, the mighty oak was once a nut like you"? It's a backhanded compliment at best, but at least it implies that your future may hold greatness.

Acorns on an oak branch

A boggy muskeg

added parasite load is affecting the population of moose in some regions. The weakened ungulates are in poorer reproductive condition and more susceptible to predation. The boreal forest supports diverse aquatic life as well. Small minnows and sticklebacks make up many of the cold-water specialists, but larger native fish include walleye, pike, lake trout, whitefish, and grayling.

Doghair Stands

Old-growth forest rightfully gets a fair bit of attention, but it's also important to recognize that early successional forest is an essential component of the landscape. Even if the habitat is "unpopular," Connecticut College professor Robert Askins wrote in *The Wildlife Society Bulletin*, many species thrive in dense doghair stands of shrubby thickets and regenerating forest. According to Askins, "Species that depend on low, woody vegetation tend to be concentrated in powerline corridors, abandoned pastures, and clearcuts. In preserved areas, maintaining shrubland habitat is frequently controversial because it requires removing trees to favor vegetation associated with human disturbance."

A few bird species that can be considered powerline specialists include blue grosbeak, golden-winged warbler, prairie warbler, and yellow-breasted chat. These utility corridors aren't allowed to mature into forests for obvious safety reasons. If left undisturbed, however, these types of habitats are ephemeral. After a roughly ten- to fifteen-year stretch, the plot's structure changes and the species that depend on it must find suitable shrub stands and young woods elsewhere.

Another early successional species is the Kirtland's warbler. This cigar-sized bird was nearly lost to the world by the mid-twentieth century. In 1961, just 502 males were detected singing in their core range of north-central Lower Michigan. Less than fifteen years later, the number plummeted to a mere 167 individuals. Kirtland's warblers are habitat specialists that thrive in thick stands of young jack pine, regenerating forest stands that had been limited on the landscape by fire suppression. Intensive forestry management practices to maintain pockets of young trees across the region have benefited this blueish-gray-and-yellow bird.

Early successional habitat, including young forests, is seasonal habitat for many species, even critters we often associate with mature woods. Ruffed grouse and American woodcock both depend on a healthy amount of young forest.

Kirtland's warbler

Branches & the Canopy

Much of what we think about when we ponder trees is the canopy—the forest's umbrella of leaves. There is a simple, innocent joy in watching a child make a leaf print by rubbing a crayon across a piece of paper on top of the backside of a leaf. And we cannot ignore the multibillion-dollar industry that revolves around fall foliage—leaf peeping. The crowns of trees create glorious canopies, be it from broadleaved deciduous or needle-bearing coniferous species. The canopy is where photosynthesis happens, keeping trees alive and releasing oxygen into the air, which in turn keeps all of us alive. (*Thank you, trees!*)

This section features icons such as redwoods and magnolias, as well as author Tristan Gooley, noted for his keen observations of the natural world and for sharing the lessons we can gather from nature. For example, krummholz trees—those that resemble flags or banners because branches don't grow on the windward side—reflect an extreme environment. Examining them helps prepare us to interpret other less obvious stories about their surroundings.

Trees are standalone beings, yet their presence influences the surrounding living and nonliving environments. In this section, we'll explore a couple types of plants, including epiphytes and mistletoes, that survive in the treetops. And finally, this section celebrates the specialized plants and animals found in the limited ecosystems of savannas—the intersection of forests and prairies.

Photosynthesis

With few exceptions, a unifying feature of plants, including trees, is the ability to produce their own food. Photosynthesis is a process by which plants convert energy from sunlight into chemical energy that they can use. Chloroplasts are the cells in plants where this energetic magic takes place. The plants gather sunlight, carbon dioxide, and water and convert them into sugars and oxygen. The sugars fuel the plants, while the oxygen is released, which allows humans and other animals to use it. And when animals consume plants, they receive a double dose of nutrients.

We perceive plants as green because the chlorophyll in plants, which is a green pigment, reflects green wavelengths while absorbing red and blue light. The light the chlorophyll molecules absorb is split into hydrogen and oxygen

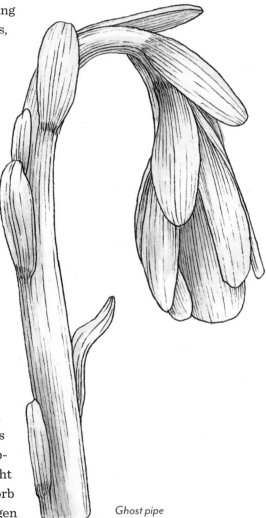

Ghost pipe

atoms. The Calvin cycle, a series of chemical reactions that produce chloroplasts, kicks in and in the process creates carbohydrates, or glucose. Water is brought up from the roots via the xylem. Stomata are tiny pores where carbon dioxide enters the plant and where oxygen exits. The resulting simple glucose molecules realign as complex starches for stored energy and cellulose, making up much of the structure of a tree.

Essentially a tree parasite, ghost pipe is an exception to the photosynthesis rule. It gathers the nutrients needed for survival by tapping into the mycorrhizal root network of fungi and tree roots. Lacking the chlorophyl typically found in plants, ghost pipe is white, almost translucent. A little bowl of a flower tops a 4- to 8-inch stem that grows individually or in small clusters.

Epiphytes

The forest canopy can provide growing perches for a variety of plants. Epiphytes, also known as air plants, are species that grow upon another plant without parasitizing the host. The host plant merely provides a perch, while the epiphytes gather all their required water and nutrients from dew, rain, and moisture from the surrounding air. Just 10 percent of plant species survive as epiphytes: ferns, bromeliads, orchids, heaths, and mosses can all survive on the trunks and branches of trees. You'll find many epiphytes thriving in warm, humid climates where there are more

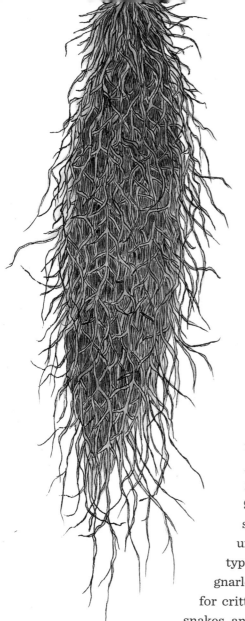

Spanish moss, a bromeliad

opportunities to drink water directly from the sky.

Some species of air plants collect water at the base of modified leaves. This botanical lifestyle is more common in tropical forests, but it is localized in northern forests as well. Spanish moss and similar air plants rely on modified scalelike filaments to gather nutrients. A double misnomer of a name, it's neither Spanish nor moss. Instead, it is one of the most widely distributed bromeliads, spanning from the southeastern United States to Central America. Its long, gray festoons adorn branches like garlands. Filaments can reach impressive lengths, more than 90 feet. These lengths are shorter segments that have naturally spliced for support. More typically, Spanish moss drapes in gnarled tangles that provide habitat for critters as diverse as frogs, bats, snakes, and jumping spiders. Following a rain shower, the strands swell as they absorb moisture and take on a greenish hue during this time of abundance. Broken strands of Spanish moss disperse, forming new colonies wherever they land.

Another epiphyte that has a limited range in southern Florida and Cuba is the ghost orchid. These beautiful flowers lack leaves; instead, filaments of root cling to crevices in tree trunks.

Needles

Gymnosperm trees are evolutionarily older than the angiosperms. Gymnosperms have seeds but no flowers, while angiosperms are what we know as deciduous, with leaves that change color and drop every autumn. With the exceptions of the ginkgoes (see One-of-a-Kind Ginkgo in Cones & Seeds), gymnosperms have leaves that are needlelike or scaled.

Nearly all conifers, including pines, spruces, junipers, yews, and others, are evergreen trees. Baldcypress and tamarack (American larch) are two deciduous conifers in North America; they shed their needles, not leaves, annually. The others all keep their needled foliage year-round. Individual needles will be lost over the years, but depending on the species, each needle can remain intact for two to forty years.

Functionally, needles perform the same roles as broad leaves. Photosynthesis takes place within the structures. Since needles remain on trees year-round, photosynthesis can occur for longer, although water becomes a limiting factor in most instances. In fact,

Douglas-fir

the conifers are more adapted to tougher climate and soil conditions. A thick, waxy coating on the needles of some species helps reduce water loss and enables them to survive frozen conditions. Needle boughs are also less prone to wind damage.

In the springtime, look for new twig and needle growth, called candles, sprouting up from the ends of branches. Examining conifer needles closely can assist with tree identification. Local species will vary, but many members of the cypress family (including cedars, junipers, and arborvitae) have scaled needles. Most true firs have a softness to the touch. Their individual needles are more flattened in cross-section, and you might not be able to roll them between your finger and thumb. Hemlock needles are similar in structure. The sharp stab of a spruce needle is an experience you won't soon forget. If you try to roll them between your fingertips, you can often feel squarelike edges on spruce needles. Pine needles grow in packets. The individual needles occur in bundles of two to five segments fused together in a single base.

Leaves

Leaves are arguably the most popular part of a tree. After all, nobody goes bark-peeping in the fall. Within the realm of trees, angiosperms are sometimes categorized together as broadleaf plants (as a contrast to the needles of the gymnosperms). Although individual leaves can vary a fair bit in size, leaf shape is one of the more reliable identifiers in plant taxonomy. Shape, however isn't always a sure thing either. The many factors that go into leaf development all help determine the results. Dryness, for instance, can shrivel up an otherwise healthy specimen. Leaves that develop closer to the tree trunk can take on

Ash tree

different structures than those farther out on the branches. Plants such as holly will respond to herbivory. That is, once animals have foraged upon a tree, it will respond by growing leaves with added spikey defenses. Leaf shape may help you narrow down the identification to a type of tree (i.e., oak or maple), but you may need to examine additional features to pin it down to an exact species.

There is much to look for when it comes to leaf anatomy. One essential characteristic to evaluate is whether they are simple or compound. Many familiar species, including maples, oaks, and aspens, have simple leaves, meaning that each leaf stem has a single blade per stalk. While simple leaves can be shaped elaborately with multiple lobes, they are still a single blade. Compound leaves have multiple leaflets per stem. Examples include ashes and many locusts. Honey mesquite, Kentucky coffeetree, palo verdes, and others have compound bipinnate leaves, which look especially feathery, with multiple subleaflets growing on multiple leaflets.

Venation in the leaves can be pinnate or palmate. In pinnate leaves, several veins branch out from a single main vein. For palmate leaves, the main veins all radiate from the base. Outer edges of leaves (the margins) can be smooth or toothed. Single-toothed species have similar-sized teeth, while double-toothed species have teeth of different sizes on the same leaf.

Leaves help define the trees they grow on and harness the light. No matter the shape they assume, leaves serve as photosynthesis engines and are essential to the life of the tree and to life on Earth.

Krummholz

Not simply a matter of trimming a few branches here and there, krummholz is an intense natural pruning process in which constant weathering results in bonsai growth. From the German words for "crooked or twisted wood," krummholz trees form in the area between the subalpine and alpine zones. Trees in the krummholz can be any number of species. For example, if we picture the treeless alpine zone as a hole, the krummholz donut around Mount Katahdin is sprinkled with balsam, fir, subalpine fir, and black spruce. Different species make up the krummholz zone out west. Subalpine fir and Engelmann spruce dominate tree growth beneath the Rocky Mountain tundra, but limber pine and lodgepole pine can also be found in some locations.

The flagging of the trees makes the prevailing wind direction obvious. What's easier to overlook is the amount of snowfall present during winter. Thicker bush growth can be an indicator of snow cover. Winter drifts help protect the branches, while exposed tree-tops take on the full force of winter weather.

Just downslope, the same species of trees grow to reach their full potential, but here in the krummholz, the stunted growth takes on more of a shrubby shape. The tallest individuals may reach only 10 feet, one-tenth of what the same species can be at lower elevations. Some trees end up growing laterally into natural swags. This twisted growth form is caused by the breaking of the apical bud, which forms at the top of the tree. These growing tips, the meristems, are one of the few parts of the tree that actively grows.

Mountaintops present challenging growing conditions for plants. Despite snow burying the area for months at a time, the environment is nearly as dry as a desert, and the wind is fierce and relentless.

Flagging on a krummholz tree

Alpine cushion plants eke out an existence in squat bunches above timberline, while the gnarliest of trees push their upper ecotone of survival in the krummholz zone. Curiously, these difficult growing conditions aren't a detriment to survival for the individual trees. Tree-ring studies show many krummholz trees are some of the longest-lived plants, including some of the most rugged old bristlecone pines (see Old, Gnarled Sentinels in Trunks & Rings).

My, What Big Leaves You Have

While the subtleties of photosynthesis (see Photosynthesis earlier in this section) can be easy to overlook, it's impossible to ignore the importance of leaves on deciduous trees. The tree that goes all in on leaves is the American sycamore. This species has the largest leaves in North America. The broad, palmate-shaped leaves can be the size of a dinner plate, which is fitting because the leaf is the food factory of the tree. Sycamore leaves resemble maples in shape somewhat. Maple leaves have deeper indentations between the lobes, really accenting the shape, while sycamores have shallower indentations, giving them more of an oval feel and adding to the massive appearance of each individual leaf. Maple leaves grow in pairs opposite each other, while sycamore leaves are arranged in an alternating pattern. Although it has been planted beyond its native range, American sycamore is an eastern species. Western (or California) and Arizona are the sycamore types of the West and Southwest.

The leaves are notable for their size, but they aren't the most curious aspect of the sycamore canopy. Each fall, golf-ball-sized growths appear to dangle like ornaments within the sycamore treetops. Well past when the trees drop their leaves, these rough seed balls persist, sometimes hanging off branches into the following spring. Eventually, these balls drop, littering the ground with tripping hazards.

After sycamore seeds take root, the plant grows quickly. The species reaches well over 100 feet in height and takes on considerable girth. Much like a snake shedding its skin, sycamore trees will often slough off chunks of bark as the tree expands. This exfoliating bark gives the trunk a mottled, abstract pattern.

Another name for sycamore is buttonwood. Button makers have historically preferred the fine-grained lumber milled from sycamores. The buttonwood also has ties to the finance industry. In what is referred to as the Buttonwood Agreement, the New York Stock Exchange was signed into existence at a gathering under a sycamore tree in 1792.

Sycamore leaves

Fall Foliage

The colors of fall are an annual treat for naturalists and casual observers alike. The mottled quilt patterns of an eastern deciduous forest reveal rich red and vibrant orange hues. Stands of birch in the upper Great Lakes region, as well as aspen in the Rocky Mountains, shimmer and quake in golden-yellow tones. Western states see muted earthen shades along the foothill forests as oak and chaparral cling to their leaves.

In summer, green leaves are rich with chlorophyll, which captures the sunlight required for photosynthesis. Following the summer solstice, daylight hours decrease, and this increasingly shorter photoperiod is trees' environmental cue to stop producing chlorophyll for the season. As that green color fades, species reveal their different shades. Yellow and orange leaves, common on hickories, cottonwoods, and poplars, are products of xanthophylls and carotenoids, like the compounds that turn carrots into various shades of ochre. These vibrant hues are present in the leaf throughout the summer but are masked by chlorophyll. The red leaves of the maples, oaks, and dogwoods are caused by anthocyanins, which only develop in the fall with cooler temperatures. The compounds in leaves transition as trees shift nutrients to their root systems for winter.

Depending on what part of the country you're in, the optimal range of time for viewing fall leaves—and capturing those stunning foliage photographs—can stretch from mid-September to early November. Capturing the peak moment of color can be a nearly impossible scavenger hunt, whether you are traveling the Green Mountain Byway of Vermont or driving to Rocky Mountain National Park in Colorado. Drier conditions can lead to a poor showing of colors in a forest or cause plants to drop leaves early. Wetter-than-normal years are also problematic. Although yearly patterns vary and depend on local weather conditions, northern regions can expect peak colors earlier than southern areas. Similarly, the fall changes start sooner at higher elevations and arrive later in the lower valleys. While most deciduous trees drop their leaves in the fall, a few species, including beech and many of the oaks, retain them over winter. The pigments in fall leaves break down, though, so late-season leaves are crinkled and brown, thanks to tannins in the cell structure.

The deciduous life strategy of dropping leaves is one solution to winter. The tradeoff is that with the falling of the leaves, trees have hundreds of open wounds, making them more vulnerable to diseases and infestations. Every contact point between a leaf and branch is a potential avenue for disease, but trees have ways to minimize this risk. Specialized cells at the contact point link the leaves to the trees. This abscission layer grows at the base of the petiole, or leaf stem. Think of

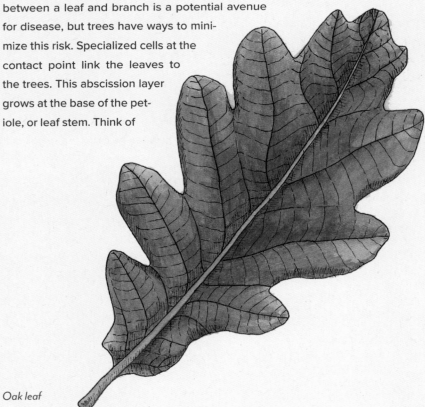

Oak leaf

this gradual thickening as a scab forming to protect the tree. By the time the leaves drop from the branches, these tree scabs are nearly healed over.

This fall, as leaves accumulate on the ground, pick up a few. Examine the palette of colors. Some leaves will be fully transformed. Others will show splotched patterns with hints of lingering green. Give those petioles a closer look. While it's amazing to think that no matter how massive a tree is, the entire flow of life-sustaining nutrients moves through these tiny leaf portals, it's even more impressive that trees have adapted a strategy to maximize this benefit in summer but shed foliage annually to prepare to be dormant in winter.

Jumping in a pile of leaves is a time-honored tradition. As leaves pile up like confetti on the ground, though, don't feel pressured to rake them all up. A growing movement to leave the leaves where they fall is gaining momentum. Gather a few of the tree discards for composting if you want, but by not hauling all the leaves away, your yard can be a refuge for several critters. Many invertebrates overwinter in the leaf litter, relying on it for food and shelter. Small mammals and songbirds sometimes use fallen leaves as nesting material. Your yard can also benefit from this natural mulch that packs a fertilizer boost and suppresses weeds. So, this year, after you enjoy nature's fall fireworks display, save your back and your backyard by leaving the leaves.

Understory

The overstory canopy gets much of the attention when we think of forests. Leaves at that upper level continue to chase the light year after year, but what is happening beneath the highest treetops?

Within any type of forest—temporal, tropical, or boreal—the understory is a unique aggregate of species. Shrubs, saplings, and herbaceous vegetation, including grasses, flowers, and ferns, dominate this zone. Vines such as poison ivy, Virginia creeper, and wild grape can also be major components.

Remember that ecological succession (see Succession in Snags & Coarse Woody Debris) doesn't always lead to the creation of forests, but when a patch of deciduous woods is in the climax stage,

Dogwood blossoms

the canopy is mostly closed. Seen from above or from below, branch spread appears nearly universal. While conifers survive in relatively poor soil conditions, deciduous trees add biomass to the forest floor each autumn in the form of fallen leaves. These added nutrients contribute to rich soils and an expansive understory, characterized by thick vegetation that creates dappled shade. Most understory plants are shade tolerant. Hollies and dogwoods thrive in the shadows of the forest. Maples and beeches can survive in the shadows, but when gaps open in the canopy, they respond quickly to this added light. Another strategy is for ephemeral plants to take advantage of light early in spring before leaf out transforms the forest. The understory growth isn't always positive, as many nonnative and invasive species have a strong foothold.

There are exceptions, but coniferous forests have a less extensive understory layer. At times, these forests are nearly barren at ground level. Sure, sunlight is minimal on the forest floor, but conifers themselves are also stunting the growth of other plants. Evergreens wear a coat of needles year-round but drop individual needles with regularity, adding tannins to the soil and preventing other plants from growing. This is also why the waters of these woods can appear tea brown. Grasses or mats of ground cover like kinnikinnick or whortleberry occasionally accentuate the sparseness of the understory. Ponderosa, longleaf, and other thick-barked pines evolved with regular, periodic ground fires, which burn the growth between the trees without harming them. These open woods can feel parklike as you wander among the trunks. Young conifer forests can be monocultures of densely packed trees fighting for limited resources; such doghair stands eventually thin out naturally as the trees mature.

Witch-Hazel Dowsing Sticks

The witch-hazel family includes eighty or so species worldwide. This group is situated right at the dividing line between shrub and tree. American (or common) witch-hazel can exist anywhere along the spectrum from single-stemmed and treelike to a dense, multistemmed thicket. Cultivated varieties (often hybrids with Asian species) are more often shrubbery. In a more naturally occurring state, the witch-hazels are understory plants.

American witch-hazel is found throughout eastern deciduous forests. Typically, it grows to about 20 feet tall, although under ideal conditions, the trunk can reach up to 35 feet in height. Most of the year, the plant blends in with the surrounding understory vegetation. In winter, however, it provides a splash of color. Bright yellow flowers emerge as the leaves are falling to the forest floor. These flowers each have four long, slender hanging petals, and they give off a rich, spicy fragrance. In a stand of woods that is mostly dormant for winter, these yellow lanterns of life offer hikers who take to the woods during the cooler season an unexpected treat. Another species, vernal witch-hazel, is a late-winter bloomer in the Ozarks and nearby Texas and Louisiana, and it transforms into its flowering glory from January to March.

Witch-hazel plants have a long history in medicine and folklore. Tinctures and extracts are used in many first aid and skincare products. The first commercial witch-hazel extraction facility was established in Connecticut in 1866, and the state continues to lead in the production of products derived from witch-hazel.

Witch-hazel

Forked twigs from these plants were also the preferred branches for witching sticks or dowsing rods. Some people believed that the branches would twitch in the presence of water, and many wells were dug with the aid of witch-hazel. The name "witch-hazel" likely derives from this practice. *Wicke* comes from Middle English and translates to "wicked," and *wych* is the Anglo-Saxon word for "bend."

Tree Silhouettes

As Tristan Gooley explains in *The Lost Art of Reading Nature's Signs,* "The wind shapes individual trees, just as it does the whole woods." The effects of wind on trees are obvious under the most extreme conditions. Krummholz (see Krummholz in Branches & the Canopy), with flagged branches pointing the same direction the prevailing winds blow, demonstrates this effect. With practice, you can notice the same effects play out in more subtle variations. Even within sheltered woodlots, the highest, most exposed branches likely reveal the direction of the prevailing winds. Gooley calls it the "wind-tunnel effect," where the windward side of a tree flattens, while the downwind branches remain straggly. Winds also indirectly alter underground roots. Trees often grow thicker roots on the upwind side to drop an anchor and secure themselves.

Wind is but one environmental consideration that alters tree growth—water also sculpts every tree. The drier the conditions, the shorter plants are likely to be. Availability of light can also affect the overall shape of a tree and the individual leaves on a branch. In habitats north of the equator, the branches on the north side of trees grow more vertically, while sunnier, southern-facing stems stretch outward. Gooley terms this the "tick effect" because these trees' profile looks kind of like a checkmark. This sharp contrast is often accentuated with more branches growing on the sunnier side.

As Gooley points out, "If we are surprised by what we find, there is usually a good explanation to be found. Randomness is not a great survival strategy, so it is rare in nature." The more we observe, the more we will notice.

A beech tree shaped in part by wind

Magnolia

As you might expect from a sweet Southerner, the magnolia is a charming tree. Robust foliage and full crowns give magnolias a pleasing essence. This dense upper level creates a lovely, shaded

Magnolia blossoms

area underneath that's perfect for a picnic. In addition to their stately beauty, the plants, especially some cultivated hybrid varieties, are noted for having strong, delicious fragrances. There are countless, diverse varieties, which means magnolias can thrive in many different growing conditions.

The most iconic of the species is the southern magnolia. This native type holds the honor of being the state flower of both Mississippi and Louisiana. Magnolia pulls double duty in Mississippi, where it is also recognized as the state tree. Louisiana, however, bestowed that title to the baldcypress tree (see Bald Knees in Roots, Buttresses & Knees).

Worldwide, there are roughly three hundred species in the *Magnolia* genus, a group named after French botanist Pierre Magnol, a prolific researcher in the late 1600s. Magnol developed an early system to group plants into related families, like the classification system Carolus Linnaeus would eventually refine. Linnaeus assigned the name *Magnolia grandiflora* as a reference to the big (*grandis*) flowers (*flora*).

Although the flowers of magnolia trees can appear quite showy, botanically they are simple structures. The lineage of magnolia plants goes back to the Cretaceous period. These flowering plants predate bees and many other insects we often associate with pollination. Instead, the magnolias relied on numerous species of beetles to transfer pollen. The thick leathery leaves and tough carpels of magnolias can withstand bites from these insects. The female carpels look like the male stamens within the flowers, which help transfer pollen by ensuring beetles are thoroughly visiting both parts. Magnolia flowers serve as a sort of bed and breakfast for beetles. The tepals, or petal-like flowers, open during the day, and at night they close around any beetle visitors. In the morning, anthers offer fresh pollen just before the tepals open and the beetles move on, which is then transferred to a new plant's stamens. This offset timing of anthers and stamens helps minimize self-pollination.

The seeds of magnolia develop with a cone-shaped bundle. By fall, the reddish or orangish seeds burst from this follicetum, or fruiting structure.

Mistletoe

While epiphytes perch innocently on the branches of trees (see Epiphytes in Branches & the Canopy), other interlopers are more sinister. Mistletoe isn't a benign occupant of the canopy but, rather, a botanical hemi-parasite. While mistletoes conduct some photosynthesis on their own, they also tap into trees, collecting water and nutrients from these botanical hosts to supplement their needs. In Greek, the scientific name of one type of mistletoe is *Phoradendron*, which means "thief of the tree." Its common name also provides insights into the plant's life history. The Anglo-Saxon word *mistel* means "dung," and *tan* is "twig"—fitting because one major way these seeds are distributed is via bird droppings.

European, or common, mistletoe is likely the original holiday kissing ornament, but because the name celebrates excrement on a stick, the gesture seems notably less romantic. Worldwide, there are roughly thirteen hundred known mistletoe species. Within the United States and Canada, thirty or so have been documented. Hawaii alone has six unique species. The plant was introduced into the United States by 1900. American mistletoe is a leafy plant with white berries. The dwarf mistletoes look like tangles of twigs, which are more common in the western United States. Most mistletoes are wildly promiscuous when it comes to their tree partnerships. Sticky seeds are transported on fur or feathers. Alternatively, dwarf mistletoe can eject seeds at up to 60 miles per hour, sometimes being carried up to 50 feet away.

Research has shown that higher levels of mistletoe in a forest correlate with greater overall diversity. Bird species that nest in cavities

Mistletoe on its host tree

Holiday Harvests

The US Forest Service is administered under the US Department of Agriculture because timber harvests are part of the agency's multiuse mission. The bidding and contracting process for logging operators can be complex, but only a simple permit is required for the harvesting of a special tree, such as Christmas trees. Annually, the USFS issues nearly a quarter-million Christmas tree permits. The fees to cut these trees range from $5 to $25, and there is a maximum of five permits per family. More than seventy-five USFS properties allow for Christmas tree harvesting, including Ocala National Forest in Florida and Nebraska National Forest near Chadron, Nebraska. Local USFS units set strict guidelines on tree harvests. Christmas tree cutting can take place only in designated areas, and there are often requirements regarding which species may be cut, as well as height and diameter guidelines. Although permits are sold through Recreation.gov, it's a great idea to visit your local Forest Service office for tips on the best places to track down the perfect holiday decoration.

Many rules apply. Lopping off the top of a bigger tree to get that classic pyramid shape is prohibited. Instead, people must cut trees at ground level.

Harvested specimens must be at least 200 feet from main roads, so be prepared to haul the tree out. Dragging something so large across the forest floor can damage the soils and leave your decoration a bit disheveled, so figure out your plan to remove it before you break out the saw. Don't forget to bring straps to secure the tree to your vehicle.

Besides being wholesome, outdoor family fun, this targeted thinning of trees can greatly benefit forest health. In some areas, holiday tree harvests are targeted to thin dense stands of young trees, so the small saplings have plenty of space to grow in future years. In other locations, species to be cut include trees that are encroaching into grassland habitats. Having a USFS permit doesn't guarantee that you will find a suitable tree, especially if you are looking for a show-room specimen. In fact, many trees may seem flawed because they are a little asymmetrical, or they may have fewer branches than you want. Just remember, imperfection is the beauty of nature.

Don't fret—you won't have to cancel Christmas if you live in an area without a nearby national forest. You are most likely within easy driving distance of a tree farm. According to the National Christmas Tree Association, more than fifteen thousand farms grow roughly 300,000 acres of evergreens.

At the end of the holiday season, many towns offer services to collect and mulch the plants. Another option is to recycle the tree by letting it decompose in your backyard and provide shelter for your feathered friends.

are three times more abundant in forests that support abundant mistletoe. Many bird species from chickadees to goshawks will nest in or on the tangles. In one Oregon study, 64 percent of Cooper's hawk nests were associated with mistletoes. Phainopeplas, sleek crested black or gray birds of the desert Southwest, consume upward of eleven hundred mistletoe berries per day during peak berry season.

Smaller wildlife species also benefit from mistletoe—in fact, some even rely on it. The plants are the only hosts for at least three species of hairstreak butterfly. The great purple hairstreak lays its eggs on American mistletoe, and when the caterpillars hatch, they feast upon it. Thicket and Johnson's hairstreaks feed on dwarf mistletoe. These caterpillars are ingeniously camouflaged as tiny mistletoe twigs.

Mistletoe is a classic of example of why we should save the forest, even if for reasons important only to humans. Extracts from the plant are used to fight colon cancer, yet twenty species are endangered.

It is easy to mistake mistletoe for witch's broom. While the former is a hemi-parasite lodged in the upper branches of a tree, the latter is a deformity of the tree itself caused by diseases, viruses, or other damaging agents.

Redwoods

The growing conditions in which redwoods thrive extend only a few hundred miles from the Siskiyou National Forest of extreme southern Oregon to Los Padres National Forest of California. The tallest trees on Earth, *Sequoia sempervirens*, sprout from seeds the same size as the ones that grow a tomato, yet redwoods grow to astonishing heights. The loftiest of these conifers tops out at more than 360 feet in height—roughly goalpost to goalpost on a football field—the heft of which is equally impressive. Wide bases more than 20 feet across support these behemoths. That's four picnic tables lined up end to end. Plants this large require very specific habitats. In addition to more than 100 inches of rain each year, the North Coast region is often cloaked in shrouds of fog. In summer, up to 40 percent of the area's moisture comes not from rain but from fog.

Relatively cool temperatures help the forests maintain wet environments suitable for growth. The redwood stands aren't monocultures of these epic trees; instead, layers of life and tremendous diversity exist in the forests. Although they are overshadowed by redwoods, other species, including Douglas-fir, western hemlock, tanoaks, and madrones, are scattered throughout. Extensive ferns, mosses, and fungi give the understory a lushness year-round. California rhododendrons sport flashes of pink in the spring. And berries, such as huckleberry, blackberry, salmonberry, and thimbleberry, ripen at staggered intervals.

Redwoods regularly survive six hundred years, thanks in part to thick bark and heavy tannins that provide strong resistance to fires and insect damage. Trees that can live up to two thousand years aren't easy to replace, and unfortunately, only 4 percent of the 2 million acres of redwoods that

Redwood tree

once thrived remain. Combined, Redwood National and State Parks account for 45 percent of the remaining protected trees.

Being spared from the saw blades alone can't save the redwood forests. Even within the state and national park boundaries, more than two hundred nonnative species have been identified, each disrupting the natural balance of the ecosystem. Not all redwoods die in vain. Their shallow roots mean that toppling is a frequent cause of the giants' demise. In these instances, the land slowly reclaims the tree's massive biomass, and new growth sprouts upon the carcasses of decay.

American Savanna

Does the word *savanna* conjure thoughts of the African plains with wildebeests roaming grasslands and giraffes plucking leaves off acacia trees? If so, it may surprise you to learn that the same habitats exist in North America, albeit, they are increasingly rare and aren't always called savannas, but instead go by many terms, including *openings*, *parklands*, *steppes*, and sometimes even simply *woods*. There isn't a consensus on the amount of canopy cover required for savannas, but it is somewhere between a grassland and a forest. Unfortunately, the tale of these ecosystems is remarkably similar everywhere. These landscapes have been altered by human development, agriculture, and shifting fire patterns.

The species composition makes the types of savannas different. Across much of the United States, a major type of savanna is dominated by oak trees of various species. Some 90 percent of West Coast and Southwest oak lands have been cleared. What remains is affected by nonnative understory plants including English hawthorn, Scotch broom, and black mustard. Conifer encroachment is also problematic for these oak savannas.

Especially rare, thanks to altered fire regimes and human encroachment, are similar oak openings that can be found in scattered midwestern locations. With fewer ground fires, maples outcompete

Oak savanna

oaks in this part of the world. Targeted restoration work is underway in many spots, and the habitat is responding positively.

Savannas aren't exclusively made up of deciduous canopy. Pine trees can grow in scattered stands, and western foothills support ponderosa pine woods. Again, years of fire suppression have added canopy that results in conditions more similar to forests. Historically, native ungulates including bison and elk grazed freely between the trees, when the areas where more open savanna stands.

Southeastern longleaf pine stands were once the most common ecosystem in the Southeast, covering roughly 90 million acres. While the trees are still a common species, only 3 percent of the longleaf pine savanna remains as an intact ecosystem.

The loss of savanna habitats extends far beyond the loss of the trees. Hundreds of species from the oak titmouse to the Karner blue butterfly and gopher tortoise are directly associated with these now-limited habitats.

Snags & Coarse Woody Debris

Long after they die, trees remain important parts of the ecosystem. They continue to provide nutritional and structural value for many forest-dwelling animals. This section showcases these valuable connections.

Living trees can be nearly hollow on the inside. That's because the living tissues of trees are near the outer edges, just under the bark. The internal decay closer to the core speeds up as the tree dies. Standing dead trees, also known as snags, are prime woodpecker habitat, although in human-dominated landscapes, these relics are often cut down to minimize the risk of them falling.

Coarse woody debris is made up of snags that are no longer standing. These fallen logs create structure along the forest floor and provide wonderful habitat for microorganisms, such as invertebrates, and much larger organisms, such as bears that sometimes rip open the logs searching for lemon-flavored ant treats.

A combination of weather and breakdown via the decomposers, including soil bacteria, fungi, and invertebrates, eventually returns the particles of a tree back to the earth as soil. This natural cycle of succession resets the pattern of growth on a landscape.

Each entry in this section is a reminder that the life cycle of a tree continues long after its death.

Cavities

In late spring, look for the light-colored wood shavings accumulating at the base of a tree. These fresh wood chips on the forest floor may reveal one of nature's whittlers. Many species of woodpecker carve out nesting cavities deep in trunks, creating their own cabin retreats. Dead trees, be it single snags or entire stands of charred timbers following a forest fire, can make for ideal nesting locations, although some woodpeckers prefer live trees. Birds often select trees with softer wood, such as aspens and birches, for their excavations—in particular, diseased trees with heart rot are usually targeted.

An ambitious pair of hairy woodpeckers can drum out a cavity in as little as two weeks. Instead of drilling directly into the trunk, hairies often nest along stout branches. Keeping the entrance hole below an angled branch helps keep flying squirrels from taking over. Larger woodpeckers, such as the hairy, prefer a cavity that is pendulum-shaped and roughly 8 to 12 inches deep. Nearly the size of crows, pileated woodpeckers require larger holes in bigger trees that can take six weeks to complete and may be 2 feet deep when finished.

Similar in appearance to hairy and downy woodpeckers, the endangered red-cockaded woodpecker is a pine specialist of the Southeast. They live in small communal groups. Territories can extend for a couple hundred acres, and groups will develop several cavities within this range. The birds roost in individual cavities overnight. The breeding male incubates the eggs, usually in the freshest cavity site. Red-cockaded woodpeckers peck out sap wells below the entrances to the holes, which is thought to promote the flow of sap, a deterrent for predators. The internal cavity chambers extend both above and below the hole—another technique for avoiding potential

Pileated woodpecker

predators. The birds seek out trees with red heart fungus and often excavate their cavities from areas where the fungus is growing inside the trees. It can take up to two years to complete each nesting hole.

Primary cavity nesters, like woodpeckers, create their own holes. Some species will use the same hole for multiple years, but in most instances, they prefer to create a new nest site each season. The abandoned cavities aren't wasted though. Secondary cavity nesters from bluebirds to barred owls will use these prefab homes. The benefits of cavity nesting range from predator protection to thermoregulation. These habitats can be scarce on the landscape, so competition for prime real estate can be intense.

Coarse Woody Debris

Even in death, trees provide life. Before falling to the ground, a dead tree is termed a *snag*. Once it is knocked from its pedestal, the lifeless trunk and limbs become known as *coarse woody debris* (or *downed woody debris*). As you walk through a forest, notice the ghosts from the past. Recently fallen trunks will stand out, but if you look closely, you'll see signs of former trees that have been almost completely recycled back into the soil. A slight rise, perhaps a shade brighter than the immediate uplands, could represent decades of decay.

The breakdown of trees plays an important role in the nutrient cycling of a forest. Many factors go into the rate of decomposition. Forests in cooler regions have slower rates of decay than those in warmer climates. Another variable is the species of tree. Although it may seem counterintuitive as we often think of deciduous trees as hardwoods, conifers tend to break down three to four times slower than similarly sized deciduous trees.

Before the logs lose their integrity, these hunks of timber add character to the forest floor. Areas with high volumes of coarse woody debris can prove impenetrable for larger animals such as elk, but these same zones can be refuges for smaller mammals, including hares and voles. These wooden highways can also provide access for agile predators, such as fishers and martens. Reptiles and amphibians use extensively decayed logs and the damp microhabitats that coarse woody debris creates. In winter, snowfall blankets the landscape, yet stacks of fallen timbers maintain pockets of space for critters. Hibernating bears sometimes use these tangles as well.

American marten

Forest Influencer

In the early 1950s, a small black bear cub captured the hearts of an entire nation. This little critter became a cultural icon and one of the original influencers. There is still some confusion as to the name of this folk hero, even though he is instantly recognizable. Let the record show, there is no "the" in Smokey Bear. When song-writers Steve Nelson and Jack Rollins wrote the popular song "Smokey the Bear" to celebrate Smokey in 1952, they added "the" to maintain the tune's rhythm.

The US Forest Service introduced Smokey Bear to the world on August 9, 1944. The original slogan was "Smokey Says—Care will prevent 9 out of 10 forest fires," but by 1947, Smokey's catchphrase became "Remember: Only YOU can prevent forest fires!"

The fictional Smokey was hugely popular from the very beginning, but in 1950, the character came to life. That spring, there was an intense wildfire in New Mexico's Capitan Mountains, and local and regional fire crews gathered to battle to burn. An orphaned bear cub was badly burnt in the blaze, but he survived. This living symbol of Smokey Bear was moved to the National Zoo in Washington, DC, where he thrived as an animal ambassador for a quarter of a century. This Smokey Bear died in 1976 and is buried at Smokey Bear Historical Park in Capitan, New Mexico.

Since Smokey Bear's peak of popularity, forest managers have realized the value of fire in many ecosystems. Complete suppression of fires over the years has built up fuel loads in some forests, contributing to especially intense burns when fires finally do occur. Prescribed fires are increasingly used to manage and maintain functioning habitats. To keep up with the times, Smokey has adopted a new mantra: "Only YOU can prevent wildfires!" It is a small but important change. While fire is a natural part of a forest's landscape, wildfires caused by humans, which are not contained in a forest, are an enormous problem.

These massive, human-caused fires mean Smokey Bear's work is not yet done. His message has become one of the longest-running public service announcements ever, and he continues to be relevant. Smokey has a huge social media presence and following. If you want to be old school and put pen to paper, Smokey Bear has a zip code entirely to himself. Fan mail can simply be addressed to Smokey Bear, Washington, DC 20252.

H Is for Humus

Litter is the undecomposed matter that has fallen to the ground, including leaves, needles, bark, and branches. As this litter layer breaks down through physical, chemical, and biological processes, it becomes known as the fermentation layer. In a more advanced state of decomposition, the humus layer is composed of the organic materials that make up soils. In academic literature, these layers (called horizons) are often shorthanded by their initials—L, F, and H—but in the woods, you can get away with calling it all soil.

The forest floor and soil are alive with fungi, invertebrates, and bacteria. A handful of dirt contains billions of microorganisms. Mites, springtails, nematodes, and rotifers make up entire often-unseen ecosystems. While deciduous forests add layers of leaves year after year, these break down relatively rapidly. Conifer lots can feature accumulations of needles and organic matter. No matter what ground you're examining, decomposition plays a role in cycling the nutrients of the land.

Deciduous forests often have thick layers of rich soils. Moder humus (or duff mull) is a combination of the recognizable plant matter and materials that have decomposed. In this layer, invertebrates and bacteria are transforming plant parts into soils. When the organic matter breaks down further and mixes in well with the mineral soil, it is called mull humus.

Coniferous woods often harbor more humus. Fallen needles dominate this undecomposed litter. Since they can be slow to break down, the process is driven primarily by fungi. The decomposition process takes longer in these woods for a few reasons, including cooler temperatures, less microbial life, more acidic soils, and the composition of the needles themselves.

Springtail

Litter Critters

Because *leaf litter* sounds like something you'd want to blow off your sidewalks and driveway, the term should be rebranded. Litter critters are a vital component of a healthy forest ecosystem. The litter, duff, and humus layers are accumulations of dead plant materials, including leaves, needles, and twigs. A component of the nutrient cycle of the woods, the decomposition process is driven by microorganisms and fungi, but this forest floor habitat is home to critters of many sizes too.

Ovenbirds, small songbirds in the warbler family, are olive brownish and have a dark-lined orange crown stripe on the top of their heads. While most warblers are noted for flitting about in the treetops, ovenbirds are a species of the forest understories of North America, from the northern Rockies to the Atlantic Ocean. Ovenbirds glean much of their diet by foraging about for invertebrates in the leaf litter. These birds are named for the remarkably complex nest structures they build directly on the ground. Working from the inside, female ovenbirds construct roofed domes over woven cup-shaped nests. The side entrances to these duff abodes resemble miniature outdoor baking ovens, hence the name. However, this covered protection isn't enough to prevent brown-headed cowbirds from laying eggs in ovenbird nests. Ground nests are also easy targets for many predators. Chipmunks commonly raid nests, as do jays, weasels, and squirrels, not to mention barred owls and red-shouldered hawks.

A healthy layer of leaf litter is also required for wood frogs to survive. These small amphibians spend the winter burrowed in the humus of the forest floor, frozen solid for more than half the year. When temperatures rise in spring, the frogs defrost and emerge from the leaf litter. While other amphibians are hibernating in the mud at the bottom of ponds, wood frogs get a jump-start on early season breeding.

Ovenbird

Nursery Logs

Most seeds never stand a chance at growing into a tree, but fallen timbers create microhabitats that provide conditions uniquely suitable for germinating seeds. These nursery logs can give seeds a leg up during their most vulnerable early days of growth.

Downed timbers can passively alter an ecosystem by changing the amount of sunlight that reaches the forest floor. A bank of seeds accumulates, waiting for conditions to be suitable for germination. A gap in the canopy can create a sharp uptick in sunlight and prompt the seeds to sprout.

A more direct way nursery logs benefit germinating seeds is by providing added nutrients. The decomposing coarse woody debris is a bountiful resource for any tiny seedlings that land upon it (or perhaps were left behind in an animal scat). Similarly, the sponginess of decaying logs or an abundance of fern and moss can provide

Seedling sheltered by a nursery log

a boost of retained moisture by slowing runoff. These slight shifts can have huge repercussions, even under seemingly similar growing conditions a few short feet away.

Tender young plant shoots are tasty morsels for countless herbivores. But thanks to nursery logs, a few seeds can avoid becoming critter snacks. The surviving trees eke out an existence under the shelter of the log. Blowdowns or windthrows can create tangles of protected pockets that give seedlings a jump-start on growing up.

Disease is another problem young trees sometimes have to cope with. Soil diseases can often inhibit seedling germination and growth, and nursery logs can serve as a buffer while the young plants establish themselves and build up a resistance to the pathogens.

On your next nature stroll, see if you can spot little baby trees that are trying to make a go of it under the protective watch of a nursery tree.

Petrified Forest

Situated east of Holbrook, Arizona, Petrified Forest National Park has entrance stations off Interstate 40 and US Highway 180. Many travelers find the park a worthwhile detour, and a driving tour will only spark your curiosity. Plan to spend a few hours, days, even weeks exploring the landscape. The visual reminders of a planet you can hardly recognize will inspire us to protect the earth as we look ahead to the next 200 million years.

Plant a Tree

Pop quiz: Do you know which state is the birthplace of Arbor Day? You may be surprised to learn that it's Nebraska. Planting trees is great, but it's essential to recognize that forests aren't the only biomes in the world. Grasslands are the climax community for much of the Midwest and Great Plains, including vast stretches of Nebraska, the original home of Arbor Day.

The tree "holiday" was the brainchild of political appointee and Nebraska newspaper editor J. Sterling Morton. (Morton's son founded the Salt Company bearing the same name.) The first Arbor Day was celebrated in 1872 and approximately 1 million trees were estimated to have been planted during the inaugural festivities. The idea spread to all fifty states and around the globe. Most locations celebrate Arbor Day on the last Friday in April, although other locales have shifted the celebration to a more appropriate time for planting. In 1972, after one hundred years of the holiday, the National Arbor Day Foundation was officially established. The nonprofit organization is headquartered at the Morton property near the banks of the Missouri River in Nebraska City.

The fifty-plus-room Morton mansion, now called Arbor Lodge, is open for tours year-round. Some visitors may prefer to wander around the 260-acre property. Grounds include an impressive collection of orchards and plantings, plus a network of nature trails along the forested bottoms of Table Creek. A visitor favorite is Treetop Village, an elevated network of netted walkways and unique bridges that connects eleven different treehouses and covers 3 acres. The WonderNet feels a little like a trampoline in the sky, and the 50-foot slide is a fun option for exiting the canopy. It's not all fun and games at Arbor Day Farm and the Arbor Lodge

State Historical Park though. The sustainably certified Lied Lodge hosts many environmental meetings and conferences.

Of course, Arbor Day is the foundation's most famous event, but the Tree City program is also widely recognized. This initiative kicked off in 1976, and nearly half a century later, more than 3600 communities are certified as tree cities, recognizing their commitment to maintaining and expanding healthy trees in their communities.

Perhaps more impressive than the cumulative payoff of Arbor Day programs are the less tangible results of raising the collective consciousness around the importance of trees. Anyone can plant a tree. As the phrase goes, "The best time to plant a tree is twenty or thirty years ago. The second-best time is now."

Spruce seedling

Even with relics of fossilized trees scattered across the windswept grasslands, it's hard to imagine what this landscape once looked like. This region of eastern Arizona has sparse tree cover. A few junipers cling to life, sometimes growing directly upon the petrified timbers of the past. But in the Triassic period (227 to 205 million years ago), this dry region was humid and tropical.

Fed by many tributaries, an expansive river basin flowed through the area during the Triassic period. The riparian zone was lush with trees, ferns, and giant horsetails. Uplands had 200-foot-tall conifers standing alongside ginkgoes and cycads. Giant reptiles, amphibians, and insects lived with early dinosaurs. Fish thrived in the flowing waters.

The fossil record within the park is extensive. In addition to the namesake forest fossils, many Ornithodirans (birds and their extinct relatives) and Pseudosuchians (crocodilians and their extinct relatives) have been discovered here. After being quickly engulfed by sediment 200 million years ago, fallen timbers fossilized over hundreds of thousands of years. Minerals slowly replaced the porous wood of the buried logs. Most petrified wood in the region now consists of quartz, including clear and smokey quartz, purple amethysts, and yellow citrine. As the Colorado Plateau experienced gradual uplifting, the fragile quartz-infused trees broke into distinct segments. Every piece of petrified wood serves as a giant crystal reflecting a kaleidoscope of colors, thanks to quartz coupled with iron, carbon, and manganese impurities.

Logs transformed into quartz

Succession

Ecologically, succession is the process by which habitats change over time. Broadly, there are two types: primary and secondary. Each represents the shifting of vegetation in response to environmental changes. Early successional specialists thrive in open sunlight and are quick to move in following a disturbance. As growing conditions change, a different set of plants will be better suited for survival. The climax community is the apex of plants expected and represents a stabilization in the range of species found under the local conditions; however, don't think of it as the pinnacle of the land. It's merely one extreme along the spectrum.

Primary succession occurs on newly exposed surfaces. One source of new earth is volcanic activity—fresh lava is an ecological blank slate. Similarly, shifting dunes and retreating glaciers expose soils, kicking off the primary succession process. In the years and decades following the eruption of Mount St. Helens in Washington State, succession has played out along the slopes.

Fireweed

More common, secondary succession is like a reset on the process. Fire and blowdowns are examples of the disruption of a climax forest. It can take decades for the land to return to forest. Human behavior can also reset, or in some cases stunt, successional processes.

It's easy to think of succession as a linear process of plant communities transitioning from one cohort to another. But as with all things in nature, the reality is much more dynamic. A mosaic of ecosystems blankets the natural landscape.

Henry Chandler Cowles of the University of Chicago championed the concept of succession. Much of his observation and work took place at the dunes along the southern shores of Lake Michigan, culminating in his 1898 dissertation *An Ecological Study of the Sand Dune Flora of Northern Indiana*. Visitors to Indiana Dunes National Park can take a hike along the Dunes Succession Trail to see this process at various stages.

Succession doesn't turn on and off—instead it is constantly fluctuating. While the species change from place to place, the interplay between plants and environmental conditions is universal.

Don't Move Firewood

Local plants are usually resistant to invasive species within their native range, but once invasive species establish a foothold, they can spread quickly. The "Don't Move Firewood" campaign was developed in 2008 to bring awareness to how easily nonnative forest insects and diseases can be spread. Regulations vary by location, but in general, it is recommended to not transport firewood more than ten miles away. When camping, it is best to find a local source of firewood for your s'mores station. This simple act can help stop the spread of nonnative invasive pests.

Emerald ash borer has become the face of invasive pest management, but this is just one of many problematic species wreaking havoc in North American forests. Other examples include Asian longhorn beetles, hemlock woolly adelgids, and the moth *Lymantria dispar*. Many of these insects are imported in shipping materials and lumber or as hitchhikers on ornamental plantings.

Emerald ash borer

The emerald ash borer (EAB) is less than half an inch long, yet the species has been nearly single-handedly wiping out ash trees since it arrived in the Detroit, Michigan, and Windsor, Ontario, region in the mid-1990s. The insects lay their eggs on ash bark. After hatching, larvae forage about under the bark, leaving behind winding trenches. Metallic-looking green adults leave behind D-shaped exit marks as they move to nearby trees and repeat the infestation cycle. To date, the EAB has been found in thirty-five states and five provinces.

There are native Cerambycid beetles in North America, but the Asian longhorn evolved in China and the Korean Peninsula. Unlike EAB, which targets ash trees, Asian longhorn beetles damage maples, birches, willows, buckeye, elms, and others. This species' distribution continues to expand in eastern North America after it was first discovered in New York in 1996.

Hemlock woolly adelgids are aphid relatives. Some species are native to the Pacific Northwest, but a type that originated in Japan was discovered in Virginia in 1951. For thirty years it infested mostly ornamental hemlocks. By the 1980s, it began spreading widely in natural forests. Now it has been documented in seventeen states and is killing hemlock trees by the thousands.

In addition to ecological harm, there are serious economic ramifications to nonnative species. The *Lymantria dispar* moth (and other foliage-eating pests) causes more than $869 million in damages in the United States each year. Local governments spend nearly $2 billion annually on tree removal, replacement, and treatments.

Menacing Manchineel

Few people in North America recognize the manchineel. The species, which is familiar throughout Mexico and the Caribbean, extends into southern Florida. It would behoove you to learn how to identify this plant, especially if you plan to spend any time in this part of the world. While most trees are beloved, the manchineel strikes fear into many people's minds. In 2011, *Guinness World Records* proclaimed it the most dangerous tree in the world.

The manchineel's tempting green, bite-sized, applelike fruits can help identify it. On trees without fruits, which can reach up to 40 feet in height, look for reddish-gray bark. The tree often grows intermingled within mangrove trees. In parts of the Caribbean, the easiest way to identify the toxic plant is to look for the red X or band painted on the bark as a warning.

What makes manchineel such a terrifying encounter? Basically everything. The milky sap of this poinsettia relative packs a punch. Cutting the tree is dangerous enough, but burning the plant releases noxious vapors into the air. Contact with skin causes it to burn.

Manchineel tree

Spaniard Juan Ponce de León likely died from manchineel poisoning at the hands of a Calusa arrowhead dipped in the tree's sap.

The black-spined iguana doesn't understand what the fuss is about regarding the manchineel. Immune to the toxins, the reptile enjoys loafing on the tree limbs as well as snacking on the dangerous fruits.

Ghost Forests

Habitats are a dynamic spectrum, and the successional process is in a near-constant state of flux (see Succession, earlier in this section). The climax community should not be considered the pinnacle or the most successful habitat. Ecologically, functionality depends on a mosaic of successional states.

One region where habitat is noticeably transitioning is along coastal salt marshes. Stands of trees eke out an existence along the brackish interface between fresh and salt water. While these trees are tolerant of relatively high levels of salinity, they cannot survive in salt water. A one-two punch of droughts followed by hurricanes can speed up the demise of the coastal forest in some instances. The parched landscape may take on a surge of salt water, which ultimately kills off entire forests. The standing snags can remain prominent, ghostly gray sentinels for years as the vegetation and waterscape change around them.

Snags in a coastal salt marsh

This shift in conditions can be ideal for the nonnative Phragmites—reed grasses—to move in and form a monoculture of dense, reedy habitat. In other instances, the opening of the canopy allows salt-tolerant shrubs and salt marsh grasses to thrive. These healthy marsh ecosystems support a different array of species than the forests they replaced. For species such as various woodpeckers and the prothonotary warbler, these ghost forest snags become new homes. Woodpeckers carve out nesting cavities in the snags, but prothonotaries also benefit as they are one of only two species of warblers that nest in hollow tree cavities. Marsh specialists also thrive in this new interface between land and sea. Secretive by nature, seaside sparrows, clapper rails, and least bitterns slink about in the salt marshes left behind.

While coastal forests once supported many bird species, losing this sliver of habitat creates more opportunities for another community of plants and animals in its wake.

Forest Ecology

Ecology highlights the relationships between living organisms and their environments. The forest biome—the plants and animals that naturally occur in the forest—is full of fascinating interactions. As the first professor of wildlife management, Aldo Leopold, wrote in *A Sand County Almanac*, "That land is a community is the basic concept of ecology, but that land is to be loved and respected is an extension of ethics."

This section begins with an entry that isn't about trees at all. Instead, it showcases forests of different sorts. Neither Joshua trees nor saguaro cacti are considered trees botanically, yet stands of these plants are forests in structure and function from an ecological framework. Other non-tree features of the forest are also discussed, from spring flowers that rely on ants for pollination, to a classic keystone forest species: the beaver.

Land management practices can crop up in our own backyards or across entire governmental agencies. One example touched on shares the very hands-on approach to the forests at El Yunque National Forest in Puerto Rico. Although El Yunque is one of the smallest units within the US Forest Service, it is home to several threatened and endangered plant and animal species.

We will also look at the human connections to forests and trees. As Leopold observed, "A land ethic, then, reflects the existence of an ecological conscience, and this in turn reflects a conviction of individual responsibility for the health of land."

The final entry highlighting Leopold's concept of "Thinking Like a Mountain" may perhaps inspire current and future generations to cultivate and live by their ecological consciousness.

Forests Without Trees

There are many definitions and parameters of the word "forest," but a forest is technically a collection of trees over a large area. How many trees does it take to make a forest? Stands of closely packed trees are obvious forests in the basic sense, although the forest threshold is vaguer when examining savannas. The Food and Agriculture Organization of the United Nations stipulates the trees be at least 5 meters (16 feet) tall and cover an area of at least 0.5 hectares (12 acres), and that the canopy covers at least 10 percent of the area. While trees are a requirement for true forests, there are a couple of ecosystems similar to forests that lack any timbers.

A Joshua tree forest is perhaps a misnomer. Joshua trees aren't botanically trees, but ecologically these stately stands of yuccas fulfill many of the same functions as forests. The tallest Joshua trees can tower 40 feet—quite a lot taller than most other desert plants. The

Saguaro cactus

branching arms of the plants replicate a leafy canopy for the Mojave Desert zone. And the various layers of the bayonet structures each appeal to different wildlife species. In the desert, where shade is a limited commodity, Joshua tree trunks provide microclimates of relief.

Similarly, saguaro cacti fill forest niches in the Sonoran Desert below 4,000 feet in elevation. These slow-growing giants can approach 50 feet in height and spend decades developing their familiar arms. In Saguaro National Park, it can take thirty-five to fifty years for arms to sprout from the trunk, but in drier locations it can take twice as long. As with trees, dead snags and hollow cavities are excavated by what we could term saguaropeckers. Gila woodpeckers and gilded flickers are primary cavity nesters in the region, and their holes are used later by secondary cavity nesters such as elf owls. Harris's and red-tailed hawks will nest on platforms they build upon the crowns and trunks of saguaros just as readily as they will in proper trees.

Even if we don't see the Joshua tree and saguaros as trees, they at least play the role of honorary forests. Both swaths of the Joshua tree and stands of saguaro cacti represent the climax ecosystem for their growing zones. It is impossible to deny that spending time within these old-growth treeless forests feels like a walk in the woods.

Appalachian Trail

There is a place where you can hike among the trees for thousands of miles. A 2,200-mile stretch of trail links Springer Mountain,

Georgia, with Mount Katahdin, Maine. The Appalachian National Scenic Trail has captivated many minds since its inception in 1921. Forester Benton MacKaye came up with the initial concept. The first segments opened by 1923, and in 1925 the Appalachian Trail Conservancy was established. The official route has taken a few detours over the years, but now there is a push to extend the southern terminus to Birmingham, Alabama, via the Pinhoti National Recreation Trail—a designation that will take an act of Congress.

As you can imagine, any trail that traverses sixteen states, covers eight national forests, and passes through six National Park Service units includes widely variable habitats. Yet, you are rarely far from trees when hiking along the Appalachian Trail. Swaths of eastern deciduous forest stretch from southern stands of oak and tulip trees to northern maple and birch woods, while groves of spruce-fir conifer forests rim the high-elevation mountains in the north. The highest point along the trail is Clingmans Dome at 6,643 feet above sea level. Thanks to the relentlessly hilly terrain, the cumulative vertical elevation gain for the route is 465,000 feet.

On average more than 3 million people take a step along the Appalachian Trail each year. The number of people with ambitions of hiking the entire route in a single season is closer to three thousand. Of these, roughly 25 percent are successful. Author Bill Bryson chronicled his AT travels in *A Walk in the Woods*, and then Robert Redford secured the rights to the story and starred in the movie version. Other notable travelers include Grandma Gatewood, who at sixty-seven was the first solo female thru-hiker, trailblazing as a lightweight backpacker in 1955.

The typical AT thru-hike can take five to seven months. Fastest known times for the AT are calculated for supported and self-sufficient hikers in both northbound and southbound treks. The quickest supported athletes to date are Karel Sabbe (2018 northbound in forty-one

The Appalachian Trail, aka the Long Green Tunnel

days, seven hours, and thirty-nine minutes) and Jennifer Pharr Davis (2011 southbound in forty-six days, eleven hours, and twenty minutes).

The thru-hiker community is tightknit. Trail names are assigned, and lifelong friendships form through the shared experience of walking the long Green Tunnel, but you don't have to hike the entire Appalachian Trail to make a memory. Spoiler alert: Bill Bryson didn't. It's not even necessary to hike the official Appalachian Trail. There are eleven National Scenic Trails, including the Pacific Crest Trail and Continental Divide Trail. Pick one and take your own walk in the woods. Move slowly and stop often. Appreciate the trees as individuals and the forest as a whole.

Forest Bathing

Sure, plants generate oxygen, but that isn't the only reason human health and forest health are linked. Intentional immersions in nature have been explored for ages, but in 1982, the Japanese Ministry of Agriculture, Forestry, and Fisheries coined this practice: shinrin-yoku, or forest bathing. The meditative experience has seen a rapid rise in participation in North America since the establishment of the Association of Nature and Forest Therapy in 2012. Doctor Suzanne Bartlett Hackenmiller, the medical advisor for the organization, wrote *The Outdoor Adventurer's Guide to Forest Bathing*. In a short video on her Integrative Initiative website, Doc Suzy, as she is known, shares numerous benefits to nature therapy. Something as simple as having a potted plant in a hospital room has been shown to speed recovery in patients.

Spending time outside has even greater mental and physical benefits. Stress hormones decrease in people after just twenty minutes in nature. Memory and attention span improve 20 percent after an hour outdoors. And here is perhaps the most surprising part: the benefits are measurable, even if the participants report disliking the experience.

From the research, it is also clear that a walk in the woods can have more health benefits than a stroll on a treadmill. Plants such as

Forest bathing

onion and garlic release phytoncides as a natural defense to ward off insects. When people inhale them, antimicrobial and antifungal compounds elevate the so-called natural killer white blood cells, which help ward off disease. This process boosts the immune response and helps fight off infection. A three-day forest-bathing experience has been demonstrated to provide a response in the immune system that lasts for thirty days. Outdoor exercise has added antioxidant benefits when compared to indoor efforts too. Similarly, blood pressure drops after spending more time outdoors.

Dr. Bartlett Hackenmiller says forest bathing is "a mindful practice in nature," where a trained guide facilitates "taking nature in through the senses." She explains that many people find it a really moving experience, and participants often end up in tears during the process.

Certified forest-bathing guides lead sessions for individuals or small groups. They are trained to facilitate a deliberate slowing down of the body and directing of the senses. Self-guided reflection in nature is also a valuable experience. The most important step in forest bathing is to step outside. Let the trees be your guide.

Mountain Zones

Factors such as temperature and precipitation regulate vegetative growth. Likewise, growth is also determined by latitude, elevation, topography, and other influences.

Latitude and temperature have an inverse correlation. Higher latitudes (closer to the poles) show cooler average temperatures. For every one degree of latitude (approximately 69 miles), temperatures shift an average of about 1°C. Minnesota and Louisiana are very different places ecologically, even though they share the Mississippi River and are less than 1,000 feet different in elevation. These changes in climate are partially driven by latitude and are reflected in the trees of the areas. You won't find cypress swamps in Minnesota, while silver maple and river birch are absent from the southern stretches of the Mississippi riparian corridor.

The slopes of mountains provide a great opportunity to see altitudinal bands. As elevation increases by 1,000 feet, the average temperature decreases by 1.5° to 2.8°C, depending on humidity and time of year. Think of a stack of donuts, where each one is slightly smaller than the one below it. The top can be a donut hole (consider it the alpine zone above the

The clear effects of bands of altitude on a mountain slope

timberline). Each layer of donuts represents an altitudinal band. The top donut supports trees that can withstand colder conditions than the foothill donuts at the bottom on the stack. This Mount Donut analogy is oversimplified, and the bands aren't uniform on both sides of the mountain. Complicating matters is the rain shadow effect, which makes it drier on the side of the mountain opposite the prevailing winds. But it is challenging to pass up the opportunity to make a mountain of donuts, especially in the name of science.

Deserts are drier than grasslands, which is reflected in the vegetation. This same concept also applies to different forest types. In the Rocky Mountains, drier foothill locations support ponderosa pine and Douglas-fir. Stands of lodgepole pine dominate mid-elevation slopes. With higher precipitation needs for survival, Engelmann spruce and subalpine fir grow at the edge of the timberline. Windswept krummholz trees stand like sentinels as the highest peaks transition to alpine areas without trees.

Ribbons of Green

Floodplains bordering flowing waters support markedly different vegetation than the surrounding uplands. This swath of growth is called the riparian zone. A boost of moisture supports trees and shrubs that cannot survive in surrounding drier environments. Riparian habitats are sometimes called *ribbons of green,* as these

Cottonwood tree

narrow bands meander across the landscape and reveal the path of surface water.

In the West, riparian zones cover roughly 1 to 4 percent of the landscape. Yet, the West is a habitat rich in diversity. In Wyoming, 61 percent of the state's terrestrial vertebrates show preference for these habitats, and more than 80 percent have been documented using the riparian zone. In the Cowboy State, riparian trees include cottonwoods, box elder, willow, and American elm. Understory plants can include chokecherry, hawthorn, buffalo berry, and several types of willow. Similarly, southwestern bosques support cottonwoods, mesquite, desert willow, and desert olive as dominate tree species.

Unfortunately, many western streams are now lined with tamarisk (salt cedar), Russian olive, Siberian elm, and other invasive plants. These nonnative species offer little suitable habitat for local wildlife. Shifting hydrology variables, thanks to prolonged drought and extensive damming, have also upset the riparian zone. Decreased seasonal flooding disturbs seedling sprouting and overall rates of regeneration.

Not surprisingly, riparian corridors are favorite places for human recreation. Many community greenways are streamside trails under the shade of the riparian canopy. Although they are wonderful spots to experience nature, be careful to tread lightly within this delicate ribbon of life.

Temperate Tongass Rainforest

Most people think you must travel to the tropics to find a rainforest, but they are overlooking the temperate variety. Southeast Alaska provides a spectacular example of the temperate rainforest biome—Tongass National Forest, part of the largest coastal temperate rainforest on the planet.

At nearly 17 million acres, the Tongass is the largest national forest in the United States. It stretches from near Ketchikan in Southeast Alaska to beyond Glacier Bay National Park in the north. The Tongass National Forest surrounds much of the Inside Passage. Cruise ships bring more than nine hundred thousand visitors to the region each year, but it is still possible to get off the beaten path and into areas where bears and trees outnumber people. The nineteen designated wilderness areas within the Tongass encompass more than 5 million acres.

Female cone of a Douglas-fir

As you may imagine, trees in this coastal temperate rainforest grow quite large, thanks in part to huge amounts of annual precipitation. Many locations receive well over 100 inches of rain, and feet of snow are also common in parts of the Tongass. This moisture supports towering Sitka spruce, western red cedar, and western hemlocks. The cool summer temperatures and high volumes of rain create conditions where decomposition stalls, creating pockets of muskeg habitat. This peat bog covers 10 percent of Southeast Alaska.

The waterways of the Tongass are home to five species of Pacific salmon, fish that connect the saltwater seas to the land. After hatching in freshwater streams, young salmon move to the ocean for much of their lives. As adults, salmon use the Earth's magnetic field and olfactory clues to return to their natal waters before they breed and die. Their carcasses provide a bounty of nutrients, not only for wildlife such as bald eagles and bears but also for plants in the area.

Busy Beaver

A few species have the capacity to transform an ecosystem, even when their numbers in an area are low. These keystone animals (named after the essential block at the top of an archway that holds everything else in place) often alter the surrounding habitats. The beaver is one of the most iconic of these ecological engineers.

Beavers are well suited for gnawing down trees since they are members of the rodent family. A beaver's incisor teeth grow continually throughout its lifetime, but structurally the teeth wear down into sharp chisels. Because a strong orange enamel covers the front surface of their front teeth and the back is softer dentin, the incisors are self-sharpened with every nibble. Strong molars aid in grinding the stiff wood fibers.

Beavers are fueled by the nutritious cambium layer of the tree just under the bark. They can also consume thin twigs, leaves, and grasses, but the semiaquatic mammals don't scarf large logs, nor do they eat fish. Beavers are especially active in fall as they prepare for winter, particularly caching branches that will nourish them until spring.

Preferred tree species include aspen, cottonwood, alder, poplar, and willow. Beaver saliva contains a tannin-binding protein and specialized gut microbes, which aid in the digestion of wood fibers. Research has shown that conifer trees are sometimes used

Beaver

Champion Trees & Forests

The conservation organization American Forests was established in 1875 with a goal to create healthy and resilient forests. In the 1940s, it launched the National Champion Trees Program, which was a scavenger hunt of sorts for people to find the largest trees in the country. The program standardized a formula for quantifying the size of trees that reflects not only the height but also the girth of the trunk and volume of the canopy. A point system tallies the circumference of a tree trunk in inches, the tree height in feet, and a quarter of the average crown spread in feet. For example, the champion white fir (found in Mariposa County, California) at 251 inches around, 227 feet tall, and having a crown spread of 42

White fir

feet, scores 489 points. By comparison, the title for the biggest eastern redbud (found in Fairfax County, Virginia) has 187 points.

Florida claims the prize for the most champions on record, with more than eighty top trees. While conditions are great for plant growth, what really boosts Florida's number is its unique geography, which allows the state to host plenty of species that cannot grow in other parts of the country.

While the Champion Trees Program focuses on individual specimens, the Old-Growth Forest Network (OGFN) targets preservation and protection of entire patches of forest giants. Only 5 percent of old-growth forest remains in the western United States, and just 1 percent remains in the east. As scientist, author, and organization founder Joan Maloof states on the nonprofit organization's website, "The forest is not only something to be understood. It is something to be felt." To date, OGFN has dedicated more than one hundred forests across the country with the goal of recognizing protected forests in each of the roughly 2,370 counties that can ecologically support forests.

as "surface rafts," giving the animals a lid of sorts for their winter stockpiles.

Architecturally, beavers use toppled timbers to construct dams and lodges. By damming up flowing water, beavers create ponds and wetlands—making habitat appealing to fish. Beaver lodges are marvels of construction, although not all individuals will build one. Most lodges have two exits, including an underwater route, but the chamber the beavers live in is above the water level.

Threatened & Endangered Trees

Although coastal redwoods and giant sequoias are the poster species for threatened and endangered trees, many other trees also face intense pressure, driven primarily by habitat loss. Swaths of forest have been felled to accommodate human sprawl, and many southern species have become restricted to a small geographic area where they are especially vulnerable.

Consider, for instance, the four-petal pawpaw found in pockets of Palm Beach County, Florida. The International Union for Conservation of Nature, a collaborative effort working to document and sustain natural resources, estimates that the population of this type of pawpaw has dwindled to fewer than five hundred individual trees. This species is adapted to scrub pine sand dune habitats

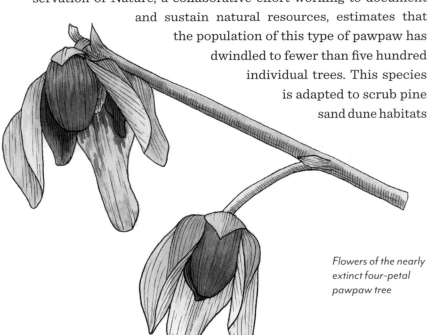

Flowers of the nearly extinct four-petal pawpaw tree

that evolved with regular fires and hurricanes. Without regular disturbance, oaks and sand pines overshadow and shade out the plant. Direct conversion of native habitats to human development also wiped out much of the pawpaw's range.

Presumed extinct for more than a decade, the Chisos Mountains oak or lateleaf oak was discovered growing in Big Bend National Park in 2022. This rarest of oaks has slim prospects, though, as a single specimen seems to be all that remains.

A few tree species will probably be lost as local conditions change. While some bird species—red-bellied woodpecker, for example—are expanding their ranges northward as temperatures climb, because trees are rooted in place, they have a limited ability to shift across the landscape. The Florida torreya, a relic of cooler climates of the past, is perhaps suited to cooler weather conditions than exist now. It may never fully recover from being hammered by fungal disease in the 1950s.

In 1994, the Boynton's sand post oak was rediscovered in Alabama. The species was presumed extinct for more than forty years when the last known Texas populations were lost. This habitat specialist lives on sandstone glades, and so it was never widespread. Species that live in narrow growing conditions are especially vulnerable to natural stochastic events and human disturbances.

One species presumed extirpated from the wilds is *Franklinia alatamaha*. Restricted to the Altamaha River Valley in Georgia, the Franklin tree (named after Benjamin Franklin) was first described in 1765 by botanists John and William Bartram. Seeds from the small tree in the tea family were sent to European arboretums and to the Bartram farm near Philadelphia. The wild populations were lost by the early 1800s. The species likely now exists only in botanical collections and arboretum plantings.

Woody Weeds

Trees are a wonderful part of the landscape—unless you are a grassland. Then these wood invaders are quite unwelcome. When they appear where they don't belong, trees just become weeds.

Woody encroachment is part of the successional process in grassland. Fire was historically the main reset button for prairie ecosystems. Long-term research from Konza Prairie in Kansas shows there is more to the story than claiming burning kills off shrubs and trees while grasses survive. Following a burn, dogwood and sumac grow

Sumac tree

both upward and outward. Burning in four-year cycles stimulates this radial growth and expands woody encroachment into grasslands. These clumps of shrubs are also believed to act as tree nurseries, since birds inevitably plant seeds as they visit.

Other types of weedy trees are the nonnatives that have taken over natural areas, many of which were planted because they grow quickly. Some, such as Russian olive, were widely planted as windbreaks, for soil stabilization, and as a berry resource. Others such as mimosa and Callery (Bradford) pear started as ornamentals but have since gone feral. These nonnative species ultimately provide native wildlife little value. They also outcompete species that support the local ecosystems. For example, species such as the tree of heaven use allelopathic chemicals to kill off native vegetation.

Nonnative species decrease an area's overall biodiversity. As author and naturalist Aldo Leopold stated, "A thing is right when it tends to preserve the integrity, stability, and beauty of the biotic community." Nonnative species disrupt all three pillars and have contributed to the decline of nearly half the threatened or endangered species in the United States. Combating these dangers unfortunately comes with a hefty multibillion-dollar price tag each year.

El Yunque National Forest

El Yunque is one of the smallest, yet most biologically diverse units of the US Forest Service. This 29,000-acre gem in Puerto Rico is the only tropical rainforest under the umbrella of the US Department of Agriculture. The land supports more than 250 different species of trees—more than all the other national forests combined.

El Yunque is a pocket of forest that hints at how Puerto Rico and the Virgin Islands looked some five hundred years ago. Roughly 33 percent of the tree species found in El Yunque are endemic to the region, meaning they aren't found anywhere else. Conversely, approximately 120 nonnative tree species have been documented in the national forest, many of which were brought to Puerto Rico initially to bolster efforts to produce lumber.

El Yunque recognizes at least six forest communities, and elevation, soil type, and precipitation all affect the species composition. The cloud forests of Puerto Rico occur above approximately 2,000 feet in elevation, where water condensation helps drive plant growth.

The forests support an array of wildlife species. One rare warbler found here, the elfin-woods warbler, was not described by scientists until 1971. Superficially, it resembles the more widespread black-and-white warbler, but the elfin-woods warbler prefers dense vines, and its nests are often tucked among epiphytes. About the time researchers documented the elfin-woods warbler, another endemic bird, the Puerto Rican parrot, was nearly extirpated from the wild. By 1975, only thirteen parrots remained in the wild, but they hung on and even expanded for another half century. In 2017, Hurricane Maria wiped out the last fifty-five wild parrots. Captive breeding efforts have allowed researchers to release birds back into the wild. El Yunque National Forest and Rio Abajo are both now home to released, endangered Puerto Rican parrots. The parrots depend on the trees for nesting, food, and shelter for protection from predators; this dependence is yet another example of how everything in the natural world is connected.

Endemic birds of El Yunque National Forest, clockwise from top left: Puerto Rican parrot, lizard cuckoo, and tody

Redwoods of the East

The largest remaining old-growth bottomland hardwood forest in the United States survives within South Carolina's Congaree National Park. The park was established in 2003, but nearly thirty years earlier, in 1976, Congress had established the area as Congaree Swamp National Monument.

Wood storks in a swamp

The park includes more than 20,000 acres of federal wilderness area. Board-walk trails weave through the forests of Congaree, but a watery wilderness kayak or canoe adventure is the best way to appreciate the giant timbers. The Congaree River Blue Trail, a 50-mile paddle, links downtown Columbia, South Carolina, to the park and was designated a National Recreation Trail in 2008.

Both the Congaree and Wateree rivers deposit nutrient-rich sediments along the floodplains, which, coupled with a long, hot growing season, helps the region support some of the tallest timbers. The area is sometimes referred to as the Redwood National Park of the East. Although the forests were harvested heavily in the early 1900s, there has been less logging since the mid-twentieth century, and the trees have recovered and thrived. Congaree National Park is noted for having the highest concentration of Champion Trees. To date, fifteen specimens have been identified as the largest of their kind. Loblolly pines tower higher than 160 feet, and cypress trees have circumferences of up to 25 feet. But despite being on protected land, the massive trees of Congaree aren't immune to threats. Keeping out nonnative species is a constant battle within the park.

Beyond the trees, the biodiversity within the park is also impressive. From synchronous fireflies to wood storks, many elusive species eke out an existence in this endangered ecosystem.

Fleeting Forest Flowers

Forests can seem barren in the depths of winter, but the first signs of spring reveal themselves slowly as flowers. Skunk cabbage is suited for wet habitats, including marshy woodlands and wet thickets from

North Carolina and Tennessee north to Nova Scotia and southern Quebec for the eastern species, and Northern California to Alaska for the western variety. These plants are often in full bloom before the snows have fully melted for the season. Skunk cabbage generates a little heat, which protects the plant from freezing. Its cabbage name is a nod to the larger leaves, and "skunk" refers to its putrid odor, a stench that attracts pollinating flies. The flowers are nearly enclosed within a blotched brownish, terrarium-like structure called a spathe.

In the deciduous forests of the east, spring ephemerals comprise an entire suite of flowers. They take advantage of the sunlight hitting

Trillium in bloom

the forest floor early in spring before leaf-out closes the canopy. Some, including trout lily, trillium, hepatica, and bloodroot, thrive in this seasonal window, thanks in part to a partnership with ants. This myrmecochory is mutually beneficial to insects and plants, since ants get food resources as they eat a nutritious oily coating called an elaiosome, and plants get their seeds distributed. Similarly, many of these spring ephemerals, such as Dutchman's breeches, depend on native bee species for pollination, while the bees rely on these earliest bloomers for nutrition too.

These insect connections make spring ephemeral plants especially vulnerable. They are limited to patchy distributions and are susceptible to threats, including habitat fragmentation, competition with invasive species, and even trampling from people wandering off-trail. You should always leave no trace when you visit the wilds, but in the woods in springtime, it's especially important to watch your step, as footprints can also cause damage.

Ferns

Originating in the Devonian period, ferns are living fossils that were abundant in the time of the dinosaurs. A few towering treelike fern species remain, but now most are small understory plants. The lushness of a forest understory full of ferns can invoke feelings of a fairyland. There is something inviting about the long, feathery leaves called fronds. The vibrant green hues add to this sense of abundance

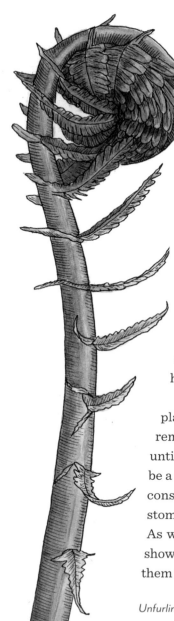

in the forest. Ironically, these monocultured stands of ferns provide surprisingly limited wildlife habitat, although epiphytic ferns can boast a bounty of biodiversity in the canopy. Ground ferns offer some amphibians shelter.

Growing conditions vary across the spectrum, but in general, each fern species has a specific set of growing conditions under which it will thrive, making them solid indicators for localized environments. In the New England forests, the endemic maidenhair fern grows in moist soils high in nutrients. Species such as fairy swords and star cloak ferns survive in exposed deserts far from the forests. Ostrich fern is cold hardy and common in many Canadian forests.

Fiddlehead ferns are a life stage of the plant—not an individual species. Young ferns remain curled up and resemble a fiddle scroll until the fronds start to unravel. Fiddleheads can be a culinary delight, but go easy if you choose to consume them. Even the edible ones can cause a stomachache if you consume them in abundance. As with most greens, a simple but tasty way to showcase fiddleheads' earthy overtones is to saute them in butter and garlic.

Unfurling fiddleheads

Cheniers

Water is a powerful force. While we often think of the erosion that flowing water can cause (think of the Grand Canyon), all this soil is ultimately deposited. Along the heel of Louisiana is a unique type of oak forest that owes its existence to twelve thousand years of sedimentation. The mighty Mississippi River carries vast amounts of fine grains of sand, silt, and clay particles. This upstream dirt settles out as a delta where the broad, slow-flowing river empties into the Gulf of Mexico. After centuries, the river meanders and shifts, and new delta deposits are dropped. These bands of accumulated soils offer forest oaks a foothold.

Called cheniers (French for "place or grove of oaks"), these slight ridges of linear oak forests run parallel to the Gulf of Mexico. They stand tall above the marshes that surround them. What were relic

Cheniers in the midst of marsh

coastal beaches are now separated from the salty waters by thin strips of marsh. The conditions that allowed the trees to grow here made cheniers vulnerable to human development. Fewer than 10,000 acres of chenier habitat remain in Louisiana. The rest—up to 500,000 acres by some estimates—has been lost to roads, agriculture, mineral extraction, and human settlement.

Cheniers serve several ecological functions along the Gulf Coast. They are a first line of defense against storms and aid in the prevention of saltwater intrusions. These isolated woodlots also serve as essential wildlife habitats for birds and butterflies. As migrants cross the Gulf of Mexico, they touch down into the first trees they reach, especially if they encounter weather systems. These fallout conditions can mean life or death for birds, and many perish. Others can find food and shelter in these chenier stands and survive to continue their northbound springtime journeys.

Thinking Like Leopold

In 1995, the National Park Service and the US Fish and Wildlife Service released fourteen wolves into Yellowstone National Park. The following year, an additional seventeen were set free. The large canines had once roamed the region, but the species was extirpated from northwest Wyoming and surrounding areas nearly seventy years earlier.

The absence of a top, or apex, predator altered the landscape drastically, and the ramifications of returning *Canis lupus* to the region continues to play out. The simplified version of wolf restoration implies that with wolves back, elk can no longer safely loaf around browsing in the riparian corridor, chomping down on all-you-can-eat tree buffets. Wolves keep the elk herds on alert and moving about, so more sapling trees have been able to survive. This increased recruitment of the forest has, in turn, been celebrated as the cause of the restoration of beaver, songbirds, and even trout. This is a trophic cascade

Wolf, an important predator whose presence improves the health of landscapes

response where the apex predator has been the primary cause for the changes. While trophic cascades have been demonstrated with many predator species, the data from Yellowstone is less clear. Organisms and effects in ecology are not isolated and linear. No doubt the ecosystems of Yellowstone have responded to the return of the wolf, yet additional factors are working synergistically.

Grizzly bears also make a living munching on the occasional elk, with calves being targeted most frequently, but bears can take down adults too. Some data shows grizzly predation has increased on young elk as a response to declining numbers of spawning cutthroat trout. Cutthroat populations took a hit when a different predator, the nonnative lake trout, was illegally stocked in Yellowstone Lake.

Hydrology conditions may also help explain changes in Yellowstone. Elk flourished during the seven decades they lived without threat of wolves. The herds hammered streamside willows and other riparian trees—food that beavers prefer. As beaver populations plummeted, they no longer performed their feats of ecological engineering. Beaver dams create ponds and slow down water flow rates. Without such measures, faster currents cut deeper stream channels, which ultimately causes the water table to drop, stressing willow plants.

Although Aldo Leopold, the pioneer for wildlife management as a field of academic study, was referencing an experience from the southwestern United States, he wrote in his classic *A Sand County Almanac* of his shift in thought regarding predators. After shelling down a wolf based on the trigger-itched assumption that a deer hunter's paradise would exclude wolves, he wrote, "We reached the old wolf in time to watch the fierce green fire dying in her eyes. I realized then, and have known ever since, that there was something new to me in those eyes—something known only to her and to the mountain."

He continued, "I now suspect that just as a deer herd lives in mortal fear of its wolves, so does a mountain live in mortal fear of its deer."

Another, less cited Leopold quote, "To keep every cog and wheel is the first precaution of intelligent tinkering," also applies to Yellowstone National Park. For decades the mountains were missing a fundamental cog. Returning wolves to the landscape helped right an ecological wrong and restored part of the natural processes in one of the few remaining places that can support this cast of wild characters.

Acknowledgments

While much of the writing process feels like a solo endeavor, creating a book is a huge team effort. I've been blessed with a whole bunch of all-stars on this project.

Heather Ray is my biggest fan. More importantly, she's my biggest inspiration.

Uwe Stender and TriadaUS Literary Agency have been involved with all ten of my books, spanning back over a decade. Someday I'd like to meet him in real life.

Thanks to Mountaineers Books for trusting me with this book, and for making it vastly better. Kate Rogers and the entire crew are so wonderful to work with. Special kudos to Alison Crabb, Jen Grable, Lori Hobkirk, and Laura Shauger. Emily Walker's illustrations are spot on and really elevate this book. I greatly appreciate her talents and efforts.

Thanks to Mallory and Sasha for keeping Wild Birds Unlimited B-town running while I put the final touches on this manuscript. Tip of the hat to T'Jacques for introducing me to the chenier ecosystem.

Pro tip: Find an accountant who not only understands the dollars and cents but can translate that into terms that you understand. For me, that guy is Jay Wilson.

Patrick Hogan is my unofficial, underpaid, and underappreciated promoter. To be effective, conservation efforts must bring folks together. Pat is always the first to extend an invitation or make an introduction for the benefit of the planet.

The Outdoor Writers Association of America has played a critical role in my communication efforts over the years. I appreciate the staff and volunteers who keep growing and adapting to the shifting landscape of nature content.

Lastly, my sincerest thanks to my friends and family for being so supportive over the years. Every social media share is a vote of confidence. Each book purchased is a boost.

Resources

BOOKS

Arno, Stephen F., and Carl Fiedler. *Douglas Fir: The Story of the West's Most Remarkable Tree*. Seattle: Mountaineers Books, 2021.

Barlow, Connie. *The Ghosts of Evolution: Nonsensical Fruit, Missing Partners, and Other Ecological Anachronisms*. New York: Basic Books, 2000.

Bartlett Hackenmiller, Suzanne. *The Outdoor Adventurer's Guide to Forest Bathing*. Falcon Guides, 2019.

Beresford-Kroeger, Diana. *To Speak for the Trees: My Life's Journey from Ancient Celtic Wisdom to a Healing Vision of the Forest*. Toronto: Random House Canada, 2019.

Bryson, Bill. *A Walk in the Woods*. New York: Anchor, 2006.

Cowles, Henry Chandler. *An Ecological Study of the Sand Dune Flora of Northern Indiana* (dissertation), University of Chicago, 1898.

Gooley, Tristan. *The Lost Art of Reading Nature's Signs*. New York: The Experiment, 2015.

Hugo, Nancy. *Trees Up Close: The Beauty of Bark, Leaves, Flowers, and Seeds*. Portland, OR: Timber Press, 2014.

Kimmerer, Robin Wall. *Braiding Sweetgrass*. Minneapolis: Milkweed Editions, 2013.

Knight, Dennis, George Jones, William Reiners, and William Romme. *Mountains and Plains: The Ecology of Wyoming Landscapes*. New Haven, CT: Yale University Press, 2014.

Koch, Melissa. *Forest Talk: How Trees Communicate*. Minneapolis: Lerner Publishing Group, 2019.

Leonardi, Cesare, and Franca Stagi. *The Architecture of Trees*. Translated by Natalie Danford. Hudson, NY: Princeton Architectural Press, 2019.

Leopold, Aldo. *A Sand County Almanac and Sketches Here and There*. Oxford: Oxford University Press, 1989.

Maloof, Joan. *Nature's Temples: The Complex World of Old-Growth Forests*. Portland, OR: Timber Press, 2016.

———. *Teaching the Trees: Lessons from the Forest*. Athens: University of Georgia Press, 2007.

———. *Treepedia: A Brief Compendium of Arboreal Lore*. Princeton, NJ: Princeton University Press, 2021.

Miller, John. *The Heart of the Forest: Why Woods Matter*. London: British Library Publishing, 2022.

National Audubon Society. *National Audubon Society Trees of North America*. New York: Knopf, 2021.

Nozedar, Adele. *The Tree Forager: 40 Extraordinary Trees and What to Do with Them*. London: Watkins Publishing, 2021.

Pearce, Fred. *A Trillion Trees*. Vancouver, BC: Greystone Books, 2022.

Powers, Richard. *The Overstory*. New York: W. W. Norton, 2018.

Preus, Margi. *Celebritrees: Historic and Famous Trees of the World*. New York: Henry Holt and Company, 2011.

Reid, John W., and Thomas E. Lovejoy. *Ever Green: Saving Big Forests to Save the Planet*. New York: W. W. Norton, 2022.

Sheldrake, Merlin. *Entangled Life: How Fungi Make Our Worlds, Change Our Minds, and Shape Our Futures*. New York: Random House, 2021.

Sibley, David Allen. *The Sibley Guide to Trees*. New York: Knopf, 2009.

Simard, Suzanne. *Finding the Mother Tree*. New York: Vintage Books, 2022.

Tallamy, Douglas. *The Nature of Oaks: The Rich Ecology of Our Most Essential Native Trees*. Portland, OR: Timber Press, 2021.

Trouet, Valerie. *Tree Story: The History of the World Written in Rings*. Baltimore: Johns Hopkins University Press, 2020.

Wohlleben, Peter. *The Hidden Life of Trees*. Vancouver, BC: Greystone Books, 2016.

Wohlleben, Peter, and Jane Billinghurst. *Forest Walking: Discovering the Trees and Woodlands of North America*. Vancouver, BC: Greystone Books, 2022.

———. *The Heartbeat of Trees*. Translated by Jane Billinghurst. Vancouver, BC: Greystone Books, 2021.

PODCASTS

Completely Arbortrary, by Casey Clapp and Alex Crowson, arbortrarypod.com
In Defense of Plants, by Matt Candeias, indefenseofplants.com/podcast
Ologies, by Alie Ward, alieward.com/ologies
Out There, by Willow Belden, outtherepodcast.com

OTHER

BeLEAF It or Not!, YouTube channel by Bill Cook and Georiga Peterson

"Black Forager," Alexis Nikole Nelson, @Alexis Nikole on TikTok and Instagram

Borealis, documentary about Canada's iconic snow forest, directed by Kevin McMahon, 2020.

Call of the Forest: The Forgotten Wisdom of Trees, documentary featuring Diana Beresford-Kroeger, TreeSpeak Films, 2016.

Cool Green Science, The Nature Conservancy, blog.nature.org/science.

Fantastic Fungi, documentary written by Mark Monroe, directed by Louie Schwartzberg, 2019.

"Forests," episode 8 of *Our Planet*, documentary collaboration between Netflix, WWF, and Silverback Films, 2019.

The Hidden Life of Trees, documentary featuring Peter Wohlleben, directed by Jöorg Adolph, 2020.

Intelligent Trees, documentary featuring Julia Dordel, Suzanne Simard, and Peter Wohlleben, directed by Julia Dordel and Guido Tölke, 2016.

Native Habitat Project, Kyle Lybarger and Jake Brown, www.nativehabitatproject.com

Simard, Suzanne. "How Trees Talks to Each Other," Ted Talk, June 2016, www.ted.com/talks/suzanne_simard_how_trees_talk_to_each_other?language=en.

Caring for Trees

If you want to apply your knowledge and appreciation of trees, consider volunteering with one or more of these organizations focused on conservation and causes related to trees and forests. You may also find helpful information through the extension office of your local university and at local botanical gardens as well.

American Conifer Society
conifersociety.org

American Forests
americanforests.org

American Forest Foundation
forestfoundation.org

Appalachian Trail Conservancy
appalachiantrail.org

Arbor Day Foundation
arborday.org

Arbor Lodge State Historical Park
arbordayfarm.org

Don't Move Firewood
dontmovefirewood.org

Integrative Initiative
integrativeinitiative.com/about

Lady Bird Johnson Wildflower Center
wildflower.org

Laboratory of Tree-Ring Research University of Arizona
ltrr.arizona.edu

Morton Arboretum
mortonarb.org

National Forest Foundation
nationalforests.org

National Park Service
nps.gov/index.htm

Northern Woodlands
northernwoodlands.org/about

Oak Savannas
oaksavannas.org/index.html

Old-Growth Forest Network
oldgrowthforest.net

One Earth
oneearth.org

Plant Amnesty
plantamnesty.org

Project Learning Tree
plt.org

Smokey Bear
smokeybear.com

US Forest Service
fs.usda.gov

Index

About the Author

Photo by Wes Lasher, Production House

A naturalist with a background in wildlife biology, **Ken Keffer** has written nine books highlighting the importance of exploring the outdoors, including *Earth Almanac* and *The Kids' Outdoor Adventure Book*, which won a National Outdoor Book Award. He is a regular contributor to *Birds & Blooms* magazine and The Nature Conservancy's *Cool Green Science*. Keffer uses nature as a classroom to teach people everyday lessons and has been honored as the Wisconsin Association for Environmental Education's Nonformal Educator of the Year. His rich diversity of experiences ranges from monitoring small mammals in Grand Teton National Park to studying flying squirrels in Southeast Alaska to catching blue crabs off the Maryland coast.

Keffer enjoys birding, floating on lazy rivers, and fly fishing. One of his favorite childhood memories is building Fort Fishy along Rock Creek at his grandparents' place, and one of his favorite adult memories is watching Bactrian camels roam the Gobi Desert. After living throughout the Midwest, he and his wife, Heather Ray, have put down roots in Bloomington, Indiana, where they own a Wild Birds Unlimited nature shop. Find out what he is up to at www.kenkeffer.net.

About the Illustrator

A naturalist, artist, and herbalist born and raised in the Pacific Northwest, **Emily Walker** has followed her curiosity and wonder for the natural world all over the American West. When she's not geeking out about plants, she enjoys foraging for mushrooms, cooking, and adventuring in the mountains with her partner, Graeme, and their dog, Huckleberry. Learn more about her and her art at fernsandfins.com.

MOUNTAINEERS BOOKS, including its two imprints, Skipstone and Braided River, is a leading publisher of quality outdoor recreation, sustainability, and conservation titles. As a 501(c)(3) nonprofit, we are committed to supporting the environmental and educational goals of our organization by providing expert information on human-powered adventure, sustainable practices at home and on the trail, and preservation of wilderness.

Our publications are made possible through the generosity of donors, and through sales of more than 700 titles on outdoor recreation, sustainable lifestyle, and conservation. To donate, purchase books, or learn more, visit us online:

MOUNTAINEERS BOOKS

1001 SW Klickitat Way, Suite 201 · Seattle, WA 98134
800-553-4453 · mbooks@mountaineersbooks.org · www.mountaineersbooks.org

An independent nonprofit publisher since 1960

Mountaineers Books is proud to support the Leave No Trace Center for Outdoor Ethics, whose mission is to promote and inspire responsible outdoor recreation through education, research, and partnerships. The Leave No Trace program is focused specifically on human-powered (nonmotorized) recreation. For more information, visit www.lnt.org.

YOU MAY ALSO LIKE: